Health, Technology and Society

Series Editors
Andrew Webster
Department of Sociology
University of York
York, UK

Sally Wyatt
Royal Netherlands Academy of Arts
Amsterdam, The Netherlands

Medicine, health care, and the wider social meaning and management of health are undergoing major changes. In part this reflects developments in science and technology, which enable new forms of diagnosis, treatment and delivery of health care. It also reflects changes in the locus of care and the social management of health. Locating technical developments in wider socio-economic and political processes, each book in the series discusses and critiques recent developments in health technologies in specific areas, drawing on a range of analyses provided by the social sciences. Some have a more theoretical focus, some a more applied focus but all draw on recent research by the authors. The series also looks toward the medium term in anticipating the likely configurations of health in advanced industrial society and does so comparatively, through exploring the globalization and internationalization of health.

More information about this series at
http://www.springer.com/series/14875

John Gardner

Rethinking the Clinical Gaze

Patient-centred Innovation in Paediatric Neurology

John Gardner
School of Social Sciences
Monash University
Melbourne, Australia

Health, Technology and Society
ISBN 978-3-319-53269-1   ISBN 978-3-319-53270-7 (eBook)
DOI 10.1007/978-3-319-53270-7

Library of Congress Control Number: 2017933947

Cover image © Classic Image / Alamy Stock Photo

Printed on acid-free paper

This Palgrave Macmillan imprint is published by Springer Nature
The registered company is Springer International Publishing AG
The registered company address is: Gewerbestrasse 11, 6330 Cham, Switzerland

Series Editors' Preface

Medicine, health care and the wider social meaning and management of health are undergoing major changes. In part, this reflects developments in science and technology, which enable new forms of diagnosis, treatment and the delivery of health care. It also reflects changes in the locus of care and burden of responsibility for health. Today, genetics, informatics, imaging and integrative technologies, such as nanotechnology, are redefining our understanding of the body, health and disease; at the same time, health is no longer simply the domain of conventional medicine, nor the clinic. The 'birth of the clinic' heralded the process through which health and illness became increasingly subject to the surveillance of medicine. Although such surveillance is more complex, sophisticated and precise as seen in the search for 'predictive medicine', it is also more provisional, uncertain and risk laden.

At the same time, social management of health itself is losing its anchorage in collective social relations and shared knowledge and practice, whether at the level of the local community or through state-funded socialised medicine. This individualisation of health is both culturally driven and state sponsored, as the promotion of 'self-care' demonstrates. The very technologies that redefine health are also the means through which this individualisation can occur – through 'e-health', diagnostic tests and the commodification of restorative tissue, such as stem cells and cloned embryos.

This series explores these processes within and beyond the conventional domain of 'the clinic', and asks whether they amount to a qualitative shift in the social ordering and value of medicine and health. Locating technical developments in wider socio-economic and political processes, each book discusses and critiques recent developments within health technologies in specific areas, drawing on a range of analyses provided by the social sciences.

The series has already published 17 books that have explored many of these issues, drawing on novel, critical and deeply informed research undertaken by their authors. In doing so, the books have shown how the boundaries between the three core dimensions that underpin the whole series – health, technology and society – are changing in fundamental ways.

This new book contributes to one of the series' themes, namely, the exploration of recent developments in health technology innovation and how these redefine the meaning of a condition and its framing and treatment in the clinic, as was seen, for example, in Mesman's book on neonatology (2008). In this volume, the boundaries and actors defining 'the clinic' are much more broadly based, reflecting the multiple clinical and other actors involved in managing and treating the condition known as 'dystonia', a neurological disorder characterised by uncontrollable muscular contraction. Gardner describes the ways in which we need to revise our understanding of the 'clinical gaze' (derived from Foucault's work) whereby not only patients but also their wider social circumstances and behaviour are included in both assessment and treatment regimes. The technology in question used to treat dystonia is deep-brain stimulation (DBS). The book examines the use of DBS in paediatric clinics and the emergence of what Gardner calls a 'proto-platform' – a diverse range of technical, spatial, professional and architectural contexts – within which DBS is adopted and performed. It shows how the adoption of DBS is anchored not simply in the perceived merits of DBS as a technology but equally importantly within a wider cultural set of judgements and expectations about behavioural norms and practices, and ways of understanding that drive patient-centric treatment.

The book makes an important contribution to our understanding of medical innovation through bringing together in a novel way ideas drawn from a Foucauldian tradition and from science and technology studies. It is written in a very engaging style, and its story is rich and informative. Readers will find this book an invaluable contribution towards our understanding of the complex relations between socio-technologies, health and social practices, and, in particular, how these are increasingly combined and layered as diagnostic and treatment 'platforms', involving multiple forms of evidence, disciplinary languages and professional practitioners. It also offers a timely critique of the turn towards and ideals of patient-centred medicine by showing how this is enacted through the platform that coordinates and choreographs the very patient/disorder relationship.

York, UK   Andrew Webster

Amsterdam, The Netherlands   Sally Wyatt

Acknowledgements

This book is the product of my PhD research that was part of the London and Brighton Translational Ethics Centre (LABTEC), funded by a Wellcome Trust Biomedical Ethics Strategic Award (086034). I must begin by thanking the Wellcome Trust for their generous financial support. During my PhD I was fortunate to have two excellent supervisors: Clare Williams (also director of LABTEC) and Steven Wainwright. Clare has been – and continues to be – especially brilliant. I cannot thank her enough.

My colleagues in Department of Sociology at the University of York have been supportive. Andrew Webster in particular has been extremely generous with his time – this book owes much to his endless optimism and encouragement and his insightful advice. I am forever grateful to my family and my partner Matthew for their unfailing support and encouragement.

Finally, I would like to thank my participants for their time, curiosity, and tolerance. They have shared a part of their professional and personal lives with me, and they have exposed their activities to the scrutiny of an outsider. They have provided me with a wonderful opportunity – I hope this book has made the most of this opportunity.

Chapter 3 of this book contains significant material that has been published in *Social Studies of Science* (Sage), as Gardner, J. (2013) 'A history of deep brain stimulation: technological innovation & the role of clinical assessment tools', 43(5): 707-728. Some material in Chapter 5 has been published in Gardner, J. & Williams, C. (2015) 'Corporal diagnostic work and diagnostic spaces: clinicians' use of space and bodies during diagnosis', *Sociology of Health & Illness* (John Wiley & Sons) 37(5):765-781. Material in Chapter 6 has been published in Gardner, J., Samuel, G. & Williams, C. (2015) 'Sociology of Low Expectations: Recalibration as Innovation Work in Biomedicine' *Science, Technology & Human Values* (Sage) 40(6): 998-1021, and in Gardner, J. & Cribb, A. (2016) 'The dispositions of things: the non-human dimension of power and ethics in patient-centred medicine', *Sociology of Health & Illness* (John Wiley & Sons), 38(7): 1043-1057. Chapter 7 contains some material that has been published in Gardner, J. (2016) 'Patient-centred medicine and the broad clinical gaze: Measuring outcomes in paediatric deep brain stimulation' *BioSocieties* (Springer), advanced publication online, 7 March 2016.

Rachael Allen has very kindly granted permission for the use of her artwork on page 6. More examples of her work can be found at www.rachaelallen.com.

Contents

Contents

About the Author

John Gardner is a Research Fellow in the School of Social Sciences at Monash University, Melbourne. His research is situated in medical sociology and science and technology studies, and he has published widely on neurostimulation, regenerative medicine, and innovation.

List of Abbreviations

ADL	Activities of daily living
Adm	Administrator
AMPS	Assessment of Motor and Process Skills
BFM	Burke-Fahn-Marsden Dystonia Rating Scale
BLNA	Breast Lymph Node Assay
CF	Clinical research fellow
COPM	Canadian Occupational Performance Measure
Dr M	Dr Martin – consultant
DBS	Deep brain stimulation
FDA	Food and Drug Administration
GMFM	Gross Motor Function Measure
GPi	Globus pallidus interna (palladium)
HTA	Health Technology Assessment
JG	John Gardner – Researcher
LVAD	Left Ventricular Assist Device
MDT	Multidisciplinary team report
Mm	Mum (patient's mother)
MRI	Magnetic resonance imaging
NHS	National Health Service (UK)
NICE	National Institute for Health and Care Excellence
Neuro	Neurologist
NPT	Normalization Process Theory
OT	Occupational therapist

PMDS	Paediatric Motor Disorder Service
Psyc	Psychologist
PT	Physiotherapist
QoL	Quality of Life
R&D	Research and Development
S&L	Speech & language therapist
STN	Subthalamic nucleus
TRL	Technology Readiness Levels
UPDRS	Unified Parkinson's Disease Rating Scale

List of Figures

1

Introduction: 'Where Great Need Meets Great Uncertainty'

Dystonia is a neurological disorder characterised by uncontrollable muscular contraction. In severe cases, this results in painful and crippling body postures. In 2003 the medical device manufacturer Medtronic gained regulatory approval to market their deep brain stimulation (DBS) technology as a therapeutic intervention for dystonia. Other, more standard interventions for managing severe dystonia are generally considered to be crude and ineffective. Medications provide some relief for a small number of sufferers, but for many they are ineffective or the side effects are intolerable. Ablative surgery, which involves carefully destroying a specific section of brain tissue, may have some therapeutic effect, but this is often short term; the surgery will need to be repeated, and as more brain tissue is incrementally destroyed, important functions will be irreversibly impaired. As the recent report *Novel Neurotechnologies* (2013) by Nuffield Council on Bioethics[1] states, dystonia is one of many neurological conditions for which there is 'a great need' for new, safer, and more effective therapies. There is a considerable hope among clinicians and sufferers, then, that

---

[1] The Nuffield Council on Bioethics can be described as an influential UK-based 'think tank' on bioethical issues.

© The Author(s) 2017

J. Gardner, *Rethinking the Clinical Gaze*, Health, Technology and Society, DOI 10.1007/978-3-319-53270-7_1

DBS will prove to be such a therapy. According to proponents, it is adjustable and reversible (unlike ablative surgery), and it has proved to be remarkably effective in managing the symptoms of other neurological disorders such as Parkinson's. Indeed, the Nuffield Council's report identifies DBS as one of several highly promising novel therapeutic innovations in neurology.

The report also highlights the 'great uncertainty' surrounding these promising innovations. Despite having undergone clinical studies as part of the regulatory approval process, there is a perceived insufficiency in data concerning the safety and clinical effectiveness of DBS for dystonia. For example, while some sufferers have experienced a dramatic reduction in dystonic movements with DBS, data on the actual functional gains experienced by patients are limited, and it is clear that some DBS recipients continue to experience marked functional difficulties in their day-to-day life (Gimeno and Lin 2016). And while the incidence of serious short-term adverse effects appears to be low, there is some uncertainty over the long-term effect of stimulating areas of the brain that may be implicated in cognition and mood. For these reasons, the Nuffield Council's report suggests that the use of the DBS technique for some disorders such as dystonia occupies a tenuous half-way point between experimental therapy at one extreme and routine clinical intervention at the other. This sense of precariousness is heightened by the high cost of providing DBS, particularly when healthcare authorities are becoming increasingly concerned with cost-effectiveness.

For clinicians who wish to provide DBS therapy as a routine clinical intervention, these uncertainties present considerable challenges. This book is based upon a study, using observations and interviews, of a specialist team of clinicians as they attempt to address a 'great clinical need' while managing such 'great uncertainties'. The team, which will be referred to as the PMDS, is based in the UK and is one of a few teams worldwide that are currently providing DBS therapy to children and young people with dystonia. This book explores several specific challenges encountered by the PMDS team members, and it explores how team members attempt to manage and overcome these challenges during their day-to-day clinical work. A premise of this book is that the team's response to these challenges is a vital part of the innovation process.

Indeed, the team represents what has come to be understood as a crucial and problematic stage in the translation of biomedical science into effective clinical therapies: the clinical adoption and implementation stage. There has been an emerging interest in the 'adoption' of new therapies and systems within 'the clinic' among policy makers and health service researchers, prompted by various concerns relating to, for example, the failed uptake of officially mandated innovations, perceived technology creep, and variable clinical service quality across healthcare providers. Actors tasked with facilitating innovation in highly promising, emerging areas of medicine (such as regenerative medicine) have also delineated 'adoption' as a key concern. The activities of the PMDS team, therefore, can provide some important insights into the specific types of challenges that may be associated with 'adoption', and how such challenges are managed during day-to-day clinical work. Hence, in this book I use the PMDS as a case study for examining innovation at the adoption stage.

The book therefore contributes to sociological understandings of the nature of innovation in healthcare. Drawing on a theoretical approach that is informed by medical sociology and Science and Technology Studies (STS), and by building upon work within innovation studies, I use the activities of the PMDS to reflect on the processes that underlie socio-technical change in healthcare. I introduce a novel theoretical concept: socio-technical *proto-platforms*, which brings to light the complex interweaving of heterogeneous social and technical elements that constitute medical innovation.

And, in doing this, I reflect on how understandings of health and illness are affected by innovation: the PMDS provides a valuable opportunity to examine how the implementation of a new medical technology (DBS) affects understandings of health and illness (dystonia) in an actual clinical setting. Indeed, the second premise of this book is that if we want to understand the social implications of new medical technologies, we need to explore the organisational forms – or proto-platforms – that emerge around them as they are adopted and implemented. Contrary to some of the more speculative commentary on the impact of novel neurotechnologies on perceptions of health and illness, this book provides an account of clinical work in which, due to the influence of *patient-centred medicine*, dystonia is enacted in multiple ways; in some practices as a 'biomedical' phenomenon, and in other practices as a 'social' concern. In light of this, and as a way of anticipating the social impact of the much-championed ideals of patient-centred medicine more generally, the book introduces the notion of *the broad clinical gaze* as a way of rethinking the clinical gaze described by Foucault. The broad clinical gaze, I suggest, represents an extension of medical power; power that is disciplining, and that has both constraining and productive effects.

This book can be seen as both an account of medical innovation in practice, and an account of the impact of the patient-centred ideals on day-to-day clinical work and perceptions of health and illness. By exploring the intersection of these two themes, the book provides a unique analysis of medically mediated socio-technical transformation. The specific argument and structure of the book is presented towards the end of this introductory chapter. Before presenting these, however, it is necessary to provide a more substantial background on DBS, the neurological disorder dystonia, and the PMDS clinical team. This will enable me to argue, in more precise terms, why the PMDS represents such a fruitful case study for a sociological study of innovation.

Deep Brain Stimulation (DBS)

DBS is a technique in which a pacemaker-like device is used to deliver constant electrical stimulation to areas deep within the brain. It is used to reduce the severity of symptoms for a range of neurological disorders. According to current models of the brain, such neurological disorders are the consequence of pathological or damaged brain structures which induce abnormal levels of neural activity. This activity disrupts normal signalling pathways within the central nervous system, leading to uncontrollable motor symptoms such as rigid and stiff muscles or shaking and flailing limbs (Montgomery and Gale 2008). The exact way in which DBS alleviates such symptoms is not known. One explanation is that by swamping the problematic brain structures with electrical noise it essentially masks the pulsating, cyclic neural activity that would otherwise trigger the motor symptoms (Montgomery and Gale 2008). DBS is not, then, a cure. It does not repair abnormal brain structures and it does not prevent the progressive degradation of brain tissue associated with some neurological disorders. So while it may be effective in enabling patients to regain some control over the movement of their body, when electrical stimulation ceases, so too does its therapeutic effect.

DBS is currently being used in several movement disorder therapies. Within several regulatory jurisdictions, including the European Union and the USA, regulatory agencies have permitted DBS therapies for Parkinson's, essential tremor, and dystonia. Generally, due to the perceived risks and the high cost of DBS, it is reserved for severe cases of these conditions where other, more usual treatments, have failed to provide adequate therapeutic relief. Parkinson's is by far the most common disorder treated with DBS: diagnoses of Parkinson's are prevalent in developed countries, it is progressive, and long-term medicinal treatment fails in up to two-thirds of patients (Marsden and Parkes 1977, 348); hence, according to the device manufacturer Medtronic, over 100,000 people worldwide are using DBS to manage the motor symptoms of Parkinson's (Medtronic 2015).

Some cases of epilepsy are being treated with DBS (Boon et al. 2007), and researchers are also exploring the ability of DBS to treat severe cases of psychiatric illnesses: in the European Union, DBS has been approved for the treatment of severe obsessive-compulsive disorder (OCD), and in both Europe and the USA, pilot trials assessing DBS as treatment for depression have recently been completed (Mayberg et al. 2005). In the past, severe cases of these motor and psychiatric disorders that failed to respond to medicine-based therapies would have been treated with ablative surgery. Proponents of DBS argue that unlike ablative surgery, DBS does not cause irreversible damage to brain structures, and adverse effects are therefore less likely, or easier to manage if they do arise (Benabid et al. 1991; Limousin et al. 1995).

Specialised technology is needed to deliver DBS. Electrical stimulation is delivered to the target areas within the brain by at least one lead. In an operation that lasts between three and six hours, a neurosurgeon

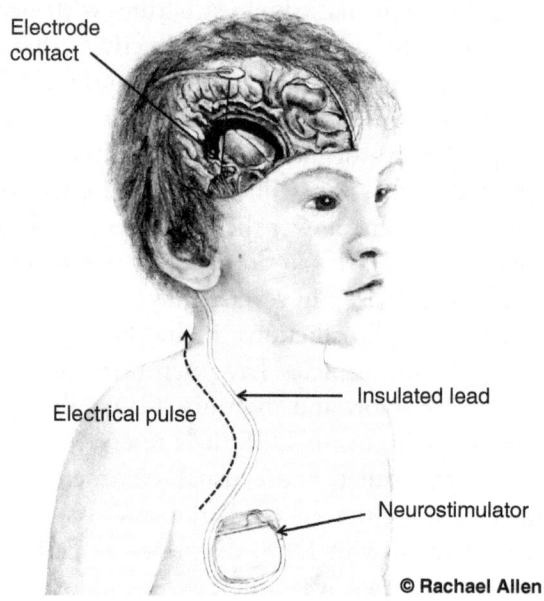

Electrode contact

Electrical pulse

Insulated lead

Neurostimulator

© Rachael Allen

Fig. 1.1 The implanted DBS system components

carefully threads the lead into the brain so that the electrodes at one end are placed within the target area. The other end of the lead is attached to a flexible extension wire, threaded under the skin from the top of the scalp, down the side of the neck to the chest (See Fig. 1.1). At this point, it is attached to a neurostimulator, also known as an implantable pulse generator (IPG), which has been implanted under the skin (Talan 2009, xiii). DBS therapies are either monopolar, where one lead is inserted on one side of the brain, or biopolar, where two or more leads are used to stimulate both sides of the brain. The specific electrode contact area within the brain depends on the condition being treated, but in motor disorders the target area is one of several structures that make up the basal ganglia region.

The neurostimulator (often referred to as an implantable pulse generator, or IPG) is very similar in appearance and function to a cardiac pacemaker, producing a constant electrical pulse, albeit at a much higher frequency than that used to regulate the beating of a heart. A battery makes up most of the bulk of the neurostimulator, and depending on the model, it may be either rechargeable or non-rechargeable. In rechargeable models the battery is recharged transcutaneously: the patient (or a caregiver) presses a recharging paddle against the area of skin that covers the neurostimulator, ideally once a day for half an hour. Non-rechargeable models tend to be used for patients who may have difficulty adhering to a recharging schedule. These batteries last between two and four years, after which the entire neurostimulator is surgically replaced.

Once the components have been implanted within the patient, a clinician can programme the neurostimulator via remote control: it can be switched on and off, the frequency of stimulation can be altered, and specific electrodes can be activated. It may take several weeks of trial and error adjustment before the most efficacious stimulation parameters for each particular patient have been determined (Ostrem and Starr 2008, 322). In some cases (particularly with Parkinson's) the patient will be given a remote programmer to enable them to have some control over the implanted device. Importantly, after the device has been implanted and the initial optimal stimulation parameters have been identified, DBS patients require long-term skilled clinical attention: six-monthly follow-up assessments are necessary to ensure that stimulation parameters are indeed still optimal (they often require fine tuning); hardware failures, such as broken or migrating leads, are not uncommon; and of course, for those patients with non-rechargeable stimulators, a neurosurgical team is needed to replace the stimulator at least every four years.

For social scientists, the emergence of DBS represents an excellent opportunity to explore the social dynamics of innovation, and indeed the tensions associated with biomedicine and society more generally. It is, as the Nuffield Council report states, considered to be a highly promising technological development. Indeed, DBS therapy for Parkinson's in particular is currently viewed as a sensible and generally effective intervention for patients who fail to respond to medicine-based therapies (Bell et al. 2009; Deuschl et al. 2006, 578). Most Parkinson's patients experience a useful reduction in the severity of their symptoms, and some patients have had dramatic improvement in their condition. Indeed, some commentators have heralded DBS as providing 'a new life for people with Parkinson's' (Chou et al. 2011), and stories of previously housebound patients with debilitating symptoms subsequently regaining independence and self-confidence with DBS are not uncommon. And, despite the high cost of DBS, there is also emerging evidence to suggest that for the long-term treatment of Parkinson's it is more cost-effective than the best alternative treatments (Bell et al. 2009; Fraix et al. 2006).

However, as with many new technological developments in medicine, DBS is linked with various risks and uncertainties. Despite the apparent success of DBS for treating motor disorders, it is associated with a range of adverse effects, some of which can be severe. There are those adverse effects that relate to the implantation of the components, such as intra-operative cerebral haemorrhaging and post-operative infection, and there are those related to the actual therapy itself, such as slurred speech and a range of so-called psycho-social effects. In DBS therapy for Parkinson's, these include mood elevation, anxiety, aggression, and cognitive dulling (Synofzik and Schlaepfer 2011, 3–4). In the longer term, some DBS patients have experienced feelings of aimlessness, despair over the years lost to their illness, marital problems resulting from mood changes, and suicide (Agid et al. 2006; Schüpbach et al. 2006). It is not clear whether these long-term effects are directly linked to electrostimulation itself or if they are the consequence of the major life changes resulting from the rapid transition from a severely impaired state to an able state (Bell et al. 2009). There are uncertainties around the frequency and severity of such effects and also around the efficacy of DBS therapies: while many patients with Parkinson's or dystonia will respond positively to DBS, some patients experience no benefit whatsoever, and identifying appropriate candidates for DBS interventions can be difficult (Schlaepfer and Fins 2010; Synofzik and Schlaepfer 2011).

The 'great promise' surrounding DBS, which is no doubt providing significant momentum to the DBS enterprise (Borup et al. 2006), has also raised concerns among commentators. As with many novel biomedical developments, DBS is subject to a fair amount of media-generated hype. The media has tended to produce 'over-optimistic portrayals' of DBS, focusing on positive outcomes and dramatic, individual cases while neglecting to explore the actual, more subtle benefits experienced by the majority of people who are undergoing treatment (Gilbert and Ovadia 2011, 2). Racine and colleagues (2007) argue that by reporting predominantly on DBS 'miracle stories', the media is creating unrealistic expectations among the public and prospective patients, and thus producing additional pressures for clinicians delivering DBS therapies. This is particularly problematic with DBS as it often represents the last hope for people with debilitating symptoms, and who have failed to improve with other available treatments (Bell et al. 2009).

Some commentators have drawn attention to the role of commercial interests in disseminating information on DBS technology. The major producer of DBS equipment is the Minneapolis-based medical device manufacturer Medtronic. According to their 2014 annual report, worldwide sales of DBS technology constituted a sizeable chunk of their USD 1.9 billion in net revenues in their neuromodulation division for the financial year (Medtronic 2014b). Fins and Schiff (2010) argue that the interplay of market forces and scientific inquiry within DBS research has resulted in potentially dangerous conflicts of interest. Medtronic, for instance, has been accused of misusing a regulatory exemption in order to facilitate the dissemination of their DBS technology into therapy for a psychiatric disorder (Fins et al. 2011). And, from a more speculative, philosophical perspective, DBS has been implicated in the 'continuous march of technologies that invade and transform the body', bringing us closer to an era of ethically contentious intelligent design, cyborgs, and mind/machine interfaces (Hester 2007, 255). Such accounts render DBS a small but definite movement towards neurotechnologies with a potential to significantly affect that which we think of as 'human' (McGee and Maguire 2007).

Thus, DBS is not only the source of great promise and hope, it is also the subject of apprehension. Indeed, it exemplifies the complex tensions associated with biomedicine in contemporary society more generally: a conviction in technology-orientated solutions, a drive to alleviate suffering, a suspicion of commercial interests, doubts over the ability of regulatory initiatives and anxiety over a precarious future. For these reasons, DBS is an excellent topic of sociological investigation. For those of us who are interested in exploring the dynamics of medical innovation, such as how the social and technological aspects of medicine may shape one another and intertwine, the relationship between developments in medicine and wider social and political context, and the ethical challenges presented by new medical technologies, DBS provides an illuminating case study.

Deep Brain Stimulation in Social Science Research

A small body of work has explored what we might call the 'social' dimensions of DBS. Some of this has been conducted by French clinical teams as part of their post-operative long-term surveillance of patients receiving DBS for Parkinson's (Agid et al. 2006; Schüpbach et al. 2006). These follow-ups have involved standardised quality-of-life and psychological assessments, along with 'qualitative psychological observations backed-up by a semi-structured psychiatric interview' (Agid et al. 2006, 410). It has been these studies that have identified the various 'psycho-social' problems encountered by some DBS recipients: marital problems, feelings of despair over the years lost to the illness, persistent negative anticipation of the future, and a loss of aim in life. Some patients, the authors note, have also experienced a sense of being dehumanised due to their reliance on a foreign, implanted mechanical technology; a feeling of being 'an electronic doll' (Agid et al. 2006, 412). Some of these same patients were interviewed by the sociologist Elsa Gisquet (2008), who highlighted the day-to-day challenges faced by people with Parkinson's. Some, she noted, lamented their reliance on a specialised medical technology, others experienced troubling changes in mood which they attributed to the therapy, and some were frustrated that DBS was not providing greater relief. Gisquet goes on to suggest that DBS therapies may pose 'a unique form of biographical disruption' for patients, and suggests that 'a period of adaption is required so that patients become used to their new capacities and accept the limits of the treatment' (Gisquet 2008, 1850).

The experiences of these patients, and clinicians' responses to them, have been more carefully interrogated by Baptiste Moutaud as part of his ethnographic study of one of the French clinical centres providing DBS (Moutaud 2011, 2015). Moutaud paid close attention to the way in which clinicians drew upon various explanatory models to make sense of their patients' responses to DBS. He notes, for example, that the clinicians were able to carefully delineate patients' psycho-social problems as being a consequence of their inability to reintegrate into 'normal life', rather than being a direct side effect of the stimulation. This enabled them to frame DBS as 'successful', and in the process they effectively drew upon and perpetuated a model of personhood characterised by a mind/body rupture (2011, 186). Patients, in other words, were understood as having a successfully managed neurological disorder, and as having a distressed mind that necessitated further psychological care. Moutaud also explored how patients themselves drew upon various models of personhood to make sense of their experiences. In contrast to their clinicians, some people with Parkinson's attributed their psycho-social problems to DBS and its direct effects on the brain. Moutaud argues (Moutaud 2011, 190) that these patients are, in effect, buying-into and reinforcing a naturalistic, neuroscientific discourse that equates behaviour and emotion with neural structures. Similarly, patients undergoing DBS for OCD (who were also included in Moutaud's study) employed biological models to understand their illness. These patients made sense of their OCD in terms of pathological brain circuits rather than, say, drawing upon psychoanalytic models that relate OCD to obsessional neuroses. Moutaud's study, then, illustrates how DBS can influence how human actors make sense of health, illness, and personhood. Indeed, like other neurotechnologies (as I discuss in Chapter 8), it has been implicated in perpetuating brain-based explanations of the self.

Moutaud's work (e.g. Moutaud 2015) also highlights other tensions. Patients, he noted, become entwined in a technologically mediated relationship with clinicians. Some experience a feeling of 'loss of control' as their DBS system often requires alteration and adjustment, and this, in turn, requires the skill and experience of clinicians. The clinical team thus had to create a care plan for patients that took this reliance into account, and that could also attend to patient dissatisfactions and potentially disabling adverse effects. This created complex service organisational patterns involving various types of staff members (neurologists, psychiatrists, psychologists, social workers, etc) that, Moutaud argues, has implications for the subsequent diffusion of the technology: the complexity and cost of the services may discourage other centres from establishing a similar service, while a slimmed-down service may be unable to adequately care for patients, thus presenting ethical concerns that could derail the technology (Moutaud 2015).

Moutaud's work, then, provides an interesting sociological insight into the transformative nature of DBS. It has influenced how people make sense of themselves and their behaviour, and it has also prompted the formation of novel organisational structures that may affect subsequence decisions about the technology's adoption. His account of DBS implementation within the clinic provides a glimpse of the complex, often unanticipated and sometimes in-harmonic socio-technical changes that constitute what we often refer to as 'innovation'. It is this transformative aspect of DBS, its capacity to influence perceptions of health and illness, and its entanglement in organisational forms that have significant repercussions for its further dissemination, that is the focus of this book. In a similar vein to Moutaud, I draw on ethnographic data to explore an organisational configuration that has emerged to deliver DBS within a particular 'pioneering' clinical setting. And, I carefully interrogate the way in which the motor disorder is rendered intelligible within this setting. In doing this, I explore how it is that a specific group of clinicians attempt to manage some of the tensions and challenges identified above: we will see, for example, how clinicians attempt to manage the expectations of patients and families (given the considerable hype surrounding DBS); and we will see how they identify which candidates are most likely to benefit from DBS. We will also see how clinicians attempt to generate evidence of the clinical effectiveness of DBS, and we examine how they coordinate themselves as a team. However, the case study of this book differs substantially from those of other social science work conducted in DBS. It is, I will illustrate, a clinical context that has been heavily influenced by the ideals of patient-centred care, and this has major implications for the transformative effects of DBS implementation.

Dystonia

The case study in this book is of a service providing DBS for children and young people with dystonia; a movement disorder that has a relatively low public profile. It is defined as 'a neurological syndrome characterized by involuntary, sustained, patterned, and often repetitive muscle contraction of opposite muscles, causing twisting movements or abnormal postures' (Jankovic 2007). In some cases it is limited to a specific region of the body ('focal' dystonia, such as cervical dystonia that affects the muscles of the neck), while in other cases it affects large sections of the body ('generalised' dystonia). In some people dystonic movements are brought on only when they attempt particular movements, while in others dystonic movements are present at rest. It is often exacerbated by stress and anxiety (Geyer and Bressman 2006). In more severe cases, the twisting movements and the resulting abnormal postures are intensely painful and severely debilitating. Although the exact pathology of dystonia is unknown, dystonia, like other movement disorders, results from abnormal neural activity in the structures that make up the basal ganglia, a region deep within the brain that is responsible for motor control. Dystonia can occur seemingly at random in people of any age.

Generally, dystonia is classified as either 'primary dystonia' or 'secondary dystonia'. Primary dystonia is used to describe cases where dystonic movement (either focal or generalised) is the only neurological condition present (Geyer and Bressman 2006). It is characterised by an absence of any identifiable brain lesion or structural abnormality, and in many cases it is associated with one of several specific genetic mutations, such as the DYT1 or DYT6 mutations (although it is not known how these specific mutations result in dystonic movements). Primary generalised dystonia is often 'early-onset', appearing in children who, up until that point, appear to have had normal motor system development. At first, dystonic movements are restricted to a particular area of the body, often the feet, before 'spreading' to other regions of the body over a period of several years until it appears to plateau (Marks et al. 2009). As this happens the child becomes increasingly disabled (especially as dystonia spreads to the trunk and upper limbs) and the child will require a great deal of care and assistance. Estimates of the prevalence of early on-set primary dystonia range from 24 to 50 cases for every million people (Defazio 2010). In this chronic form the dystonia itself is not life threatening, but it is not uncommon for such individuals to enter into a seizure-like state called *status dystonicus*. In this state, which can last anywhere from a few hours to several months, dystonic movements become so severe that life-threatening complications may ensue, such as renal and respiratory failure. Patients with *status dystonicus* often require management in an intensive care unit (Manji et al. 1998).

Secondary dystonia is used to describe cases where the dystonia results from some sort of brain injury. It is characterised, then, by the presence of a brain lesion that is rendered visible with MRI or PET imaging techniques. Depending on their exact location these brain lesions can interfere with a number of brain functions, so it is not unusual for individuals with secondary dystonia to have a number of neurological conditions such as spasticity and cognitive difficulties. A common cause of secondary dystonia is brain trauma received around the time of birth. In these individuals (who are classified as having a type of 'cerebral palsy'), dystonia will often present alongside spasticity, musculoskeletal abnormalities, and low muscle tone, leading to what clinicians refer to as a 'complex motor disorder'. Generally, secondary dystonia is not progressive. In children, cerebral palsy is the most common form of dystonia (Marks et al. 2009).

There is a subset of secondary dystonia that is progressive however. This is heredodegenerative dystonia, in which dystonia is one manifestation of a genetic abnormality that results in the slow and progressive destruction of the structures that make up the basal ganglia area of the brain. Some of these metabolic disorders are referred to as Neurodegeneration with Brain Iron Accumulation (NBIA). Here, the individual's inability to produce an essential enzymatic protein results in the build-up of particular metabolites within structures of the basal ganglia. These metabolites destroy the brain tissue, leading to a gradual increase in neurological conditions such as dystonia and spasticity until the point of death, which may occur several decades after the original diagnosis (Gregory and Hayflick 2005). Generally, as afflicted individuals progressively deteriorate and their bodies become increasingly contorted and disfigured, they experience prolonged periods of intense physical pain. These NBIA conditions are incurable. Thankfully they are very rare.

While both primary and secondary dystonia, particularly less severe cases, can be adequately managed with medicine-based therapies (such as benzodiazepines and anticholinergics), there is a subset of cases in which medicines fail, or in which the individual is unable to tolerate side effects of medication. It is this subgroup that may be considered for DBS.

At this point in time clinicians believe that the most efficacious electrode target area for managing the symptoms of dystonia is the *globus pallidus internus* (GPi) structure within the basal ganglia (Ostrem and Starr 2008).

As the Nuffield Council report states (Nuffield Council on Bioethics 2013), DBS for dystonia is considered to occupy a realm between experimental treatment and routine clinical therapy. The Dystonia Medical Research Foundation states that by 2010 around 1000 individuals worldwide had received DBS therapy for dystonia (DMRF 2010). Most of these are adults with primary dystonia, and according to the small number of reports within published literature DBS has proven to be effective in managing their condition. Some of these patients have experienced a dramatic improvement, regaining their ability to walk unaided after being wheelchair bound for several years (Ostrem and Starr 2008; Vercueil et al. 2001). As with other DBS therapies, however, there is considerable variation in improvement rates, with some patients receiving no beneficial effects whatsoever. Generally, dramatic improvements are far less likely in patients with secondary dystonia – according to the very small number of available reports. In these patients, if the dystonia itself is alleviated, the presence of additional neurological disorders such as spasticity and contractures (which do not improve with DBS) will limit the degree of functional gain experienced by the patient (Marks et al. 2009).

The Paediatric Motor Disorder Service: A 'Pioneering' Service

In 2005, the PMDS was established to provide DBS therapy to children and young people with dystonia. It was established by the current head of the service Dr Martin (pseudonym), a consultant in paediatric neurology, who had become familiar with DBS therapy for dystonia via his association with a French team providing DBS for a range of motor disorders in adults and a small number of children. During the duration of the empirical study that informs this book, the PMDS included nine team members in addition to Dr Martin, which included: an additional consultant in paediatric neurology, a clinical nurse specialist, two physiotherapists, an occupational therapist, a clinical psychologist, a speech and language therapist, a therapy assistant, and a clinical research fellow (training to become a paediatric neurologist). Aside from Dr Martin and the clinical research fellow, all members of the team are female. This 'core' team also worked closely with two neurosurgeons at a nearby hospital (where the DBS system implantation is carried out), and with several MRI and PET imaging specialists.

By the end of 2015, the PMDS had provided DBS therapy to approximately 120 children and young people with dystonia. Approximately half of this cohort are patients with secondary dystonia, a quarter have primary dystonia, and a quarter have some form of heredodegenerative disorder. In line with the predominantly adult DBS outcomes reported in the literature, the primary cases have tended to respond better than secondary cases and several have regained their capacity to walk unaided after having been wheelchair bound. DBS is offered to patients with a heredodegenerative disorder as part of palliative care: it is intended to help alleviate some of the intense pain caused by dystonic contortions, and perhaps enable the patient to regain some function. Because the PMDS is the only service of its kind within the UK, it receives referrals from around the country, including Northern Island, Wales, and Scotland. Each year the service also accepts a small number of patients from overseas.

Indeed, while there are several centres worldwide that have provided DBS therapy for paediatric motor disorders (such as the French team that originally inspired Dr Martin), the PMDS is one of very few groups that specialises in providing DBS exclusively for children and young people with dystonia. It is also the only group that has this multidisciplinary structure: the other prominent specialist team (based in the USA) does utilise the services of a physiotherapist, an occupational therapist, a neuropsychologist, and a 'child life specialist', but their specific roles differ from those of the PMDS members. The PMDS is therefore unique, and along with the US team, it is perceived to be pioneering DBS therapy for dystonia in children and young people, particularly those with secondary dystonia. As 'pioneers', team members have had to assemble new patient pathways and care plans and new protocols and practices. Team members have, in other words, engaged in various forms of *innovation work* aimed at routinising DBS for dystonia in a paediatric setting. This is within an institutional context, the National Health Service (NHS), which can provide little flexibility due to financing pressures and staffing issues. The PMDS team, which itself represents a novel multidisciplinary arrangement, has drawn upon and adapted elements from, as well as adapting to, a challenging institutional environment, to create a bespoke infrastructure for delivering DBS. It represents an example of clinical adoption, which as we will explore more fully in the following chapter, has been delineated as crucial and often problematic stage in innovation processes.

As a way of exploring this adoption of DBS in depth, this book focuses on four specific day-to-day challenges encountered by members of the PMDS team. These are challenges that resonate with the tensions relating to DBS discussed before, and they were identified during my discussions with team members and during observations of their discussions with each other. Specifically, they relate to the difficulties in coordinating a multidisciplinary team, identifying appropriate candidates for DBS, managing patient expectations, and measuring clinical outcomes. For each of these, I will draw on interviews with team members and my observations of their clinical work to illustrate how the challenge actually manifests in their clinical context, and how it is that they manage this challenge in their day-to-day clinical work. This will allow us to see that DBS has prompted a novel organisational form which has repercussions for further dissemination of DBS therapy among the paediatric neurology community. These activities of the team, as I will argue in the following chapter, can also provide us with some important theoretical insights into healthcare innovation processes. At this point, it suffices to say, a main argument of this book is that the PMDS constitutes the establishment of what can be called a *patient-centred proto-platform*, which the team members are attempting to extend into a more widely spread platform, to borrow Keating and Cambrosio's term (Keating and Cambrosio 2003). By exploring how team members manage specific challenges, I will identify some of the socio-technical elements that constitute a proto-platform.

Patient-Centred Healthcare 'In-Practice'

A key narrative in this book is that the team's activities, particularly the way in which they manage these challenges, has been informed by values commonly equated with patient-centred (or person-centred) healthcare. This is in large part a reflection of the *paediatric* focus of the service: it inhabits an environment in which, as we shall see, a deliberate effort has been made to adhere to patient-centred ideals. Indeed, the PMDS can be conceptualised as representing a confluence of neuro-technological project (DBS) and the patient-centred medicine movement. As stated earlier, a premise of this book is that if we wish to understand the social and political implications of innovations such as DBS, and indeed medical innovation more generally, then we need to understand how new technological systems are actually embedded within, adapted to, and affect concurrent healthcare movements within clinical services. For this reason, I pay particular attention to the patient-centred practices of the team throughout this book.

A patient-centred (or what is now increasingly called person-centred) approach has been championed in health sciences literature since the 1970s. In reviewing this literature, Mead and Bower (2000) have identified what they believe to be its key features. First, it endorses a *biopsychosocial perspective* (in which health and disease are seen as having clinically relevant social and psychological dimensions), as a means of overcoming the perceived shortcomings of the reductionist biomedical model of disease. Second, the approach proclaims that 'the patient' should be viewed as a *person with a unique biography* which shapes their orientation to health and illness. This orientation and the patient's own agency need to be accommodated by healthcare practices. Thus, a third feature of the patient-centred approach is its championing of 'user-empowerment', 'patient-empowerment', and 'shared decision making', in clinical contexts. Clinicians need to adopt a more egalitarian approach and recognise that their preferences may not necessarily align with those of the patient, and they need to recognise that patients' experience of disease constitutes a valuable source of expertise: patients should, therefore, be provided with the opportunity to inform clinical decision making (Mead and Bower 2000).

Thus, advocates of patient-centredness in the Europe, the USA, and elsewhere have argue that clinicians need to be responsive to the needs and viewpoints of their patients, and that the healthcare system more generally needs to accommodate the 'whole' biopsychosocial patient. The movement can be seen as a reflection of wider socio-political trends. It aligns, for example, with the valourisation of individualisation and consumerism that characterises late modernity (Thompson 2007), and it reflects a widespread rejection of the medical paternalism of the past (which has been implicated in several high-profile ethical scandals) in favour of a more 'democratic' approach (May 1992). Within many countries, key features of the approach are also being actively championed in policy discourse. Various reports in the UK on the healthcare system have proclaimed a need for 'patient-empowerment' as a way of improving healthcare outcomes and attaining greater efficiencies within healthcare systems (Department of Health 1999, 2003; NHS 2013). The NHS constitution, for example, stipulates that:

> The patient will be at the heart of everything the NHS does. It should support individuals to promote and manage their own health. NHS services must reflect, and should be coordinated around and tailored to, the needs and preferences of patients, their families and their carers (NHS 2013, 3)

Similarly, patient-centred care has been championed as a key element of care by the US Institute of Medicine (Epstein and Street 2011b).

Inevitably, there is a gap between the ideals perpetuated in policy discourse and health service literature, and actual day-to-day clinical practice. As Cribb (2011) has noted, clinical settings tend to be under considerable strain and lack the flexibility needed for translating and implementing patient-centred practices. Cribb also notes that patient-centred medicine will be operationalised in varying ways across different contexts, reflecting differences in infrastructure, expertise, and the nature of the illness being treated. Such tensions between policy and practice encourage us to question what the operationalisation of patient-centred ideals *actually looks like* in a given clinical setting, and indeed what they *should look like* in a clinical setting. We might express concern, for example, that patient-centred care could become a vehicle through which an individualistic patient-as-consumer notion of choice supplants more attentive modes of care (Mol 2008). Looking beyond the rhetoric of egalitarianism, we might ask: what spaces are provided for patient involvement, and how do these spaces configure the nature of patient involvement? And, how is patient expertise balanced with other forms of potentially contradictory expertise (such as that which is championed by evidence-based medicine) during clinical decision making? To put this in a way that aligns with the case study of this book, how are patient preferences accommodated within organisational forms that have emerged around particular biomedical projects? We may also ask: how are health and illness *actually enacted* in contexts that have been influenced by the biopsychosocial approach?

Several social science studies have sought to explore how patient-centred care is achieved in clinical settings. Dubbin and colleagues (Dubbin et al. 2013), for example, draw attention to what they refer to as the *cultural health capital* of both the clinician and the patient. Particular cultural resources and dispositions of the participants are needed to achieve an interactional dynamic that enables meaningful patient engagement in decision-making. Liberati and colleagues (Liberati et al. 2015) argue that patient-centred care also has a material dimension; tools, instruments, and other factors configure spaces that are either more or less conducive to enacting patient-centred ideals, and are thus, in some cases, as important as the participants' dispositions and tacit knowledge. The role of tools is more closely interrogated by May and colleagues (May et al. 2006) in their examination of decision-making aides used in general practice. They argue that such tools represent an attempt to create a technological solution to potential conflict between the patient involvement on the one hand, and evidence-based medicine (EBM) with its privileging of 'objective' evidence-informed decisions on the other. The tools are designed to not only reflect the current 'objective' evidence base, but they also allow for particular types of patient input. Importantly, May and colleagues point out that such patient input is highly preconfigured: patients are encouraged to proffer *particular types of information*, as prescribed by the tool: 'the patient's subjective engagement ... is reframed as a set of limited *preferences* which can be mechanically elicited, and then acted upon' (May et al. 2006, 1026 original emphasis).

In this study of the PMDS team and their attempts to manage specific innovation challenges, this book also explores patient-centred healthcare 'in-practice'. Important aspects of the environment of the team, and indeed the structure of the team itself, were designed with the intention of providing comprehensive, biopsychosocial care: as we will see, the hospital, the NHS payment structure for paediatric services, and the professions of several team members, have been influenced by patient-centred ideals. In the process of managing specific innovation challenges, team members have drawn upon these elements as well as various patient-centred tools, they have modified them to suit specific clinical aims, and they have thus created a novel, localised socio-technical infrastructure (what I will call a patient-centred proto-platform) for delivering DBS. Importantly, we explore how dystonia and the patient are *enacted*[2] within this infrastructure as team members manage each of the challenges. We pay close attention to the role of material elements in configuring the way in which dystonia is rendered intelligible, and following Dubbin and colleagues (Dubbin et al. 2013), in configuring patient involvement in decision-making. Just as May and colleagues (May et al. 2006) have noted, I illustrate that patient-centred interactions can entail eliciting particular, pre-set types of information from patients in such a way the elides 'messy' and 'cumbersome' personal detail.

This brings me to the second argument of the book. In order to delineate and interrogate the sociological and political elements of patient-centredness, I draw on Foucault's conceptualisation of the clinical gaze (1963/2003) and argue that the PMDS subjects their patients to what can be described as a *broad clinical gaze*. I define this as:

A clinical interest that extends from the shapes and structures of the body, to the subjective thoughts and emotional state of the patient, to elements of the patient's social context and their ability to act within it.

---

[2] Here I'm following Annemarie Mol's use of the term 'enactment' to replace 'social construction'. The problem with the latter, she argues, is it implies that the 'constructed' entity has obtained a degree of durability. 'Enact' conveys the short temporality of such 'constructions': a body might be constructed as a biomedical entity in one specific context, but this may have very little influence in other contexts were it is enacted differently (Mol 2002, 32–33).

I will argue that it has normative and disciplining effects, in which children and families are subjected to, compared to, and prompted towards various psychological, social, and biomedical norms. While the broad clinical gaze is not an inevitable consequence of the patient-centred movement, I will suggest that the movement represents a significant advancement in the reach of medical power.

A Summary of the Arguments and an Overview of the Structure

The following chapter provides a detailed overview of some of the social science work that has explored innovation processes in healthcare, particularly at the technology adoption stage. I begin with a discussion of theoretical understandings of innovation processes, from the now much derided but still influential conceptualisation of innovation as a linear process, to more recent understandings of innovation-as-an-emergent process. These more recent understandings highlight the important role of day-to-day clinical work in innovation processes; work that is necessary to adopt a technology within a clinical setting. I discuss some of the social science work that has explored the challenge of technology adoption in clinical settings, and in doing so, we get a sense of the types of challenges and tensions that characterise clinical contexts such as that of the PMDS. I then introduce and delineate the notion of proto-platforms, drawing heavily on Keating and Cambrosio's notion of biomedical platforms. Ultimately, the chapter provides a theoretical justification for one of the key premises of this paper: if we want to understand the socio-cultural implications of new biomedical technologies such as DBS, we need to explore the organisational forms (or platforms) within which they become embedded.

In Chapter 3, I provide an account of the historical development of DBS. By doing this, I illustrate the plurality of interests involved in the development and stabilisation of the technology, and I identify several key socio-political trends which have influenced the innovation process and that shape the environment within which the PMDS team is based. This chapter also provides a background on the DBS policy climate in the UK, and the problems this creates for the financial reimbursement for services providing DBS. This has important implications for the organisational form (or proto-platform) of the PMDS.

The subsequent four chapters each focus on one of the challenges encountered by PMDS team members. In each, we see how PMDS team members have drawn upon various tools and resources to manage the challenge, and we therefore see how a particular socio-technical element of the PMDS patient-centred proto-platform has been collectively created. Chapter 4, 'Multidisciplinary Teamwork', explores the first challenge: coordinating multidisciplinary teamwork. There has been a trend within healthcare policy to encourage multidisciplinary service provision. Yet, as various commentators have illustrated, overcoming disciplinary and professional boundaries is challenging. In this chapter I illustrate how the team manages such challenges. Specifically, I explore the role of the built environment of the hospital, their use of a collective diary, and their use of regular team meetings. It is here that I introduce and define more fully the notion of the broad clinical gaze, specific aspects of which are examined in the following chapters. Chapter 5, 'Body Work and Space', explores the challenge of identifying suitable candidates for DBS. I illustrate that for the PMDS this challenge involves differentiating dystonia (which does respond to DBS) from other manifestations of neurological pathology, such as spasticity and muscle weakness (which do not respond to DBS). In order to do this the team relies on what I refer to as the body work of the physiotherapists, particularly *communicative body work*, and *sensorial reflexivity*. Such work, I argue, requires the careful construction of an ensemble involving bodies, objects, and texts, and is an important part of the innovation process. Indeed, the embodied knowledge of the clinician, in which a particular perspectival orientation to the world has been sedimented, is an important element of the PMDS proto-platform, and platforms more generally. Within the PMDS, such embodied knowledge contributes to the enactment of a broad clinical gaze.

Chapter 6, 'Managing Expectations, Aligning Futures', explores how the team attempts to manage the expectations of patients and supporting family members, many of whom arrive at the PMDS with unrealistic expectations about what DBS can do for them. I illustrate how, in an attempt to overcome this challenge, team members use a goal-setting session with patients and supporting family members. This goal-setting session is based on the use of a specific patient-centred tool adopted from occupational therapy which delegates particular roles to team members, the patient, and family members. In doing this, it encodes particular power relations. Importantly, I argue, such tools encourage patients and families to adopt future-orientated dispositions that are conducive to technological innovation projects. I argue that these future-orientated tools, and future-management work more generally, are powerful components of the PMDS proto-platform, and indeed biomedical platforms more generally. Within the PMDS, they participate in perpetuating the broad clinical gaze.

In Chapter 7, 'Measuring Clinical Outcomes', I explore the challenge of generating 'robust evidence' on the safety and efficacy of DBS. For the PMDS team, measuring the impact of DBS is particularly challenging due to what team members feel is the inadequacy of the widely accepted clinical assessment tool in neurology for measuring dystonia. In order to overcome this challenge the team has adopted what they believe to be a more appropriate tool, a patient-centred tool, from occupational therapy. I argue that such tools are constructed perceptual systems that create useful clinical knowledge by rendering patients intelligible according to *differences of degree*. We observe the PMDS team members as they conduct an assessment with the patient-centred tool which, we see, draws the clinician's gaze to what I refer to as the patient's *domestic body technique*. This requires the careful construction of a 'domestic' assessment space within the hospital. The team's adoption of such tools reflects a general shift in many areas of clinical practice towards the attempted measurement of the 'social' aspects of illness.

In the final chapter I emphasise that the team's strategies for managing these challenges can be seen as innovation work in its own right. I also briefly illustrate some of the ways in which the PMDS is attempting to facilitate wider acceptance of DBS among the paediatric neurology community. In doing this, we glimpse some of the innovation work that is required to transform a proto-platform into a more widely spread platform. I then discuss and reflect upon some of the observations of the PMDS and the broad clinical gaze in more depth. First, I suggest that the patient-centred medicine and the broad clinical gaze can be seen as an extension of disciplinary medical power that has both constraining and productive effects. Various health policy initiatives which aim to facilitate patient-centred care may consolidate this extension of medical power. In light of this, I make a normative distinction between totalising and non-totalising disciplinary practices, and suggest that we should encourage patient-centred practices that approximate that latter. Second, I use our observations of the PMDS as an opportunity to reflect on the social effects of new neurotechnologies. I suggest that it is important not to overestimate the power of medical technologies and that we should see them as possessing interpretive flexibility. This interpretive flexibility, I suggest, constitutes a space for bioethicists and social scientists to help facilitate responsible implementation and dissemination of technologies such as DBS.

This book, then, takes as its focus the organisational form that has emerged to deliver DBS in a paediatric context, and by reflecting on this form, it ends with some suggestions on the role of social scientists in facilitating what we might call 'responsible' innovation. Throughout the book, interview data with clinicians and ethnographic field notes of interactions involving clinicians, patients, and their supporting family members are used to illustrate various points. However, readers may note that, while glimpses of the patient voice and that of their family members do come through in some of the ethnographic data, there are no data from direct interviews with patients and their family members. This reflects the original objective of the fieldwork – to explore how clinicians themselves perceive and manage challenges in innovation. So, while I had informal conversations with patients and families, I did not conduct formal interviews with them. For those readers who are interested in the more general experiences and reflections of families on dystonia and DBS, I recommend they explore the work of Allana Austin (Austin et al. 2016; Austin 2015), a clinical psychologist who undertook her PhD research into the parental experiences of secondary dystonia and DBS shortly after I completed mine.

2

Understanding Innovation and the Problem of Technology Adoption

Innovation is a key area of inquiry in economics, management and organisation studies, marketing, development studies, and of course sociology and STS.[1] In much of this theorising, 'innovation' is generally used to refer to one of three levels of activity: the generation of new technologies and products (which may be described as 'inventions'); the generation of new processes and routines; and the generation of new organisational forms. As Hill (2010) argues, it is sociology (and I would add STS) that is particularly well suited to exploring these types of activities. The core insight of sociology and STS – that behaviour and action are fundamentally shaped by social context – draws the analysts' gaze to the co-configuring interactions among these three levels (Hill 2010). It encourages us to explain, in other words, the complex ways in which inventions, new processes, and new organisational forms configure one another, and it encourages us to see 'inventiveness' and 'innovativeness' not as intrinsic qualities, but as being qualities that are actively invoked by agents within particular contexts (Barry 2001, 212). It also

---

[1] Greenhalgh and colleagues (Greenhalgh et al. 2004) have provided a useful systematic review of these disciplines and their various perspectives on innovation.

© The Author(s) 2017
J. Gardner, *Rethinking the Clinical Gaze*, Health, Technology and Society, DOI 10.1007/978-3-319-53270-7_2

encourages us to explore how innovation activities are both shaped by, and come to shape, wider socio-cultural understandings and practices, and in doing so, we are enabled to glimpse the transformative nature of innovation activities beyond their impact on markets and economic growth.

It is this broadly sociological frame that guides this exploration of the DBS and the PMDS. A premise of this book is that if we want to comprehensively explore the cultural impact of new technologies such as DBS, then we need to examine very closely how they are creatively and pragmatically implemented with actual clinical practices (i.e. organisational processes), and thus how they become embedded within wider socio-technical systems such as those that constitute the PMDS (i.e. organisational forms). In this chapter, I set out a theoretical framework that will enable us to tease out this relationship more precisely. Towards the end of this chapter, I argue that the PMDS represents what I will define as a *proto-platform* that is constituted by various socio-technical elements: architectural forms, instruments, combinations of techniques, tacit and embodied knowledge, ways of understanding, and tools for aligning visions of the future. Many of these socio-technical elements reflect values that are commonly associated with patient-centred medicine, and in the subsequent chapters, we see how these socio-technical elements participate in the enactment of the *broad clinical gaze* which tends to render health, illness, and the patient intelligible in particular ways.

The proto-platform of the PMDS has emerged as healthcare professionals have sought to implement and routinise the use of DBS within paediatric neurology. It represents, in other words, an instance of what is often referred to as *technology adoption*. In this chapter, I argue that technology adoption can be seen as an important part of the innovation process: it requires that health professionals creatively and pragmatically transform a promising technology or procedure into a workable and working process within a clinical service. I argue that this 'routine innovation' work is often elided by prevalent understandings of medical innovation, particularly the linear conceptualisation of innovation which I outline in the following section. Hence, as a way of introducing the key tenets of the proto-platform, I then outline a conceptualisation of innovation that I argue is thoroughly sociological and which adheres to the key tenets of STS: the innovation-as-an-emergent-process conceptualisation. This foregrounds the important role of the clinic (i.e. the hospital) in innovation, and it directs our analytical gaze at the practices of technology adoption and the transformative work of 'pioneering' teams such as the PMDS. We also explore some of the social science work that has studied technology adoption, as this illustrates the types of day-to-day pressures and tensions that characterise healthcare settings such as that of the PMDS, and it illustrates the types of innovation work that is needed to implement new technologies such as DBS (and hence the type of work that generates new organisational processes and organisation forms).

The chapter, then, provides a theoretical framework for making sense of the PMDS, their activities, and the potential cultural impact of their activities. In doing so, I also provide a theoretical justification for my focus on this particular clinical team. In concise terms, the PMDS provides a useful case study to explore how, generally, new biomedical technologies can prompt organisational and social change; and more specifically, how it is that the values generally associated with patient-centred medicine are becoming implicated in such organisational and social change.

The Linear Model of Medical Innovation

The linear model of innovation has received sustained critical attention, and in some respects it has become a straw man for innovation scholars. However, at the risk of flogging a dead horse, I also will begin with a critical discussion of the linear model. I do this for two reasons. First, it serves as a useful heuristic device: the obvious deficiencies of the linear model provide a useful analytical point of departure for articulating a more valuable understanding of innovation in biomedicine. And second, despite its much maligned status, it is still, according to some commentators, reflected in over-simplistic understandings of innovation held among some policy circles and also members of the biomedical community; understandings which can have detrimental effects. These include, I suggest, the elision of the important day-to-day 'clinical adoption' innovation work of clinicians working at the coalface of biomedicine.

We can characterise the linear model of medical innovation as follows: biomedical scientists have a novel idea which they incorporate into some sort of invention. This elementary invention is subject to various experimentation and testing, first in the lab under highly controlled conditions during which it is subject to various incremental modifications, and then, eventually, it is tested in select populations of humans. If the now fully developed invention (i.e. a drug or a device) sufficiently works and is reasonably safe, it will be 'placed on the market', it will be marketed, and it will be adopted into clinical practice as a new therapy or diagnostic procedure (Gelijns and Rosenberg 1994). The appeal of this model of medical innovation is that it is uncomplicated and it gives the impression of progression. It permits us to think of innovation as a process of passing through distinct, clearly demarcated stages, and, hence, as a process that can be monitored and subject to various forms of management: milestones can be identified, accountability and accolade can be attributed, grants can be awarded, hurdles can be delineated and neutralised. An obvious manifestation of this is the numerical Technology Readiness Levels (TRL) system for measuring the maturity level of an innovation, in which TRL 1 refers to 'basic principles observed and reported' and (at the other end of the scale) TRL 9 refers to an actual system that 'has been thoroughly tested in its operational environment' (Mankins 1995). The system was first developed in the 1980s by NASA; it has since been adopted into a range of technological domains in various countries as a way of planning, monitoring, and managing innovation[2] – it is often used by funding agencies to demarcate their funding priorities, for example. In the process, as actors orientate their activities according to the system, the system itself comes to partially shape innovation processes and linearity is, to some degree, enacted. Regulatory systems in healthcare also reflect the linear model of innovation, and these too can have feedback effects as biomedical innovators align their activities with regulations and guidelines. An obvious example

---

[2] The TRL system has been used in various initiatives aimed at scrutinising and facilitating innovation within regenerative medicine in the UK (UK Research Councils 2012), and the UK Government's 'Catapult' innovation-funding agencies target their support to projects at TRL 4–6.

of this is the clinical trials process itself: manufacturers of a new drug or device are required to submit particular data packages before moving from one trial phase to another. Various biomedical science policy initiatives under the banner of 'translational medicine', which has become a powerful organisational strategy in many countries, also tend to perform this linear model of innovation (Mittra 2016; Mittra and Milne 2013). Such initiatives, for example, are often framed as 'accelerating bench to bedside' innovation, and as bridging the so-called valley of death; the envisaged gulf between biomedical laboratory-based research and patient care in the clinic.

It would be wrong to declare that the linear model of medical innovation is an incorrect representation of innovation in medicine. Like any model, it creates intelligibility by foregrounding some attributes of a complex process at the expense of other attributes, and like any widely deployed model, it can have feedback effects on such processes. The worth of the model, then, is not simply a matter of its accuracy, but rather, does it do what we need it to? More specifically, the worth of a model of innovation should be judged according to the following: (1) in the process of making innovation intelligible, does it elide actors or actions which, for specific reasons, should be accentuated? (2) And, does it have feedback effects which are, for specific reasons, undesirable? While the linear model is undoubtedly of strategic value to various actors, the answer to both of these questions is 'yes': It obscures activities of important innovation actors, particularly those involved in technology adoption, and by obscuring these it has – at least according to some commentators – undesirable consequences.

The linear model gives the impression that useful therapeutic and diagnostic advancements ultimately have their genesis in basic biomedical research. One consequence of this is that biomedical research has tended to receive vast amounts of funding from governments and agencies, while 'other important activities, like clinical and health services research have been largely neglected' (Morlacchi and Nelson 2011, 512). This is problematic because if we reconceptualise medical innovation (as I do shortly), we can see that considerable innovation can occur at the site of clinical delivery and in health services contexts. Indeed, the linear model gives the impression that the clinical adoption stage is relatively unproblematic; it is as if the new drug, device, or technology is by this stage fully formed and ready for use, and can be adopted by clinicians in a manner that is predictable and largely uniform. The model, then, does not account for the various ways in which the local specificities of a clinical context influence how a drug, device, or technology is used (or indeed whether it is used), and while it may presume that healthcare professionals will need to learn how to use the technology, it does not account for this 'learning-in-practice' as being a generator of novelty: learning-in-practice as a source of innovation in its own right is elided (Morlacchi and Nelson 2011). It constitutes, in other words, a form of what some STS theorists have referred to as 'invisible work' within medical innovation (Star and Strauss 1999). The linear model of innovation also lends itself to the fallacy of what economists have referred to as demand pull. Here the innovation process is perceived to be driven by and orientated towards a pre-identified market demand for, say, a therapeutic intervention for a particular illness. The transformative implications of medical innovation processes, such as their capacity to influence professional and institutional structures within medicine, and their capacity to shape understandings of disease, illness, and the body, are obscured (see, e.g. Joyce 2006; Pasveer 1989; Martin 1999). As de Laat has argued, markets are the result of the innovation enterprises, rather than an *a priori* determinant (De Laat 2000, 190). The linear model, in other words, ignores the important point that medical innovation may entail a constant reconceptualisation of the problem being addressed, and may involve an ongoing negotiation about what constitutes a satisfactory endpoint. As Brown and colleagues (Brown et al. 2000) have demonstrated, prospective endpoints are actively constructed and often contested during innovation processes.

The critical shortcomings of the linear model, then, are that it obscures the complex processes that constitute clinical adoption (processes which can be significantly innovative in themselves) and that it obscures what we might call the co-construction entailed in innovation; the way in which innovation processes are both shaped by, and shape the social, professional, and institutional context in which they take place. In particular, it draws the analytical gaze (and research funding) away from the collectively constructed *novel organisational forms* that can emerge in the process of implementing a new technology. My task for the remainder of this chapter, then, is to put forward an alternative conceptualisation which foregrounds these particular aspects of innovation. Such a conceptualisation is necessary to support 1) my focus on the PMDS as an example of an *organisational form* that has emerged to implement a therapy in a particular setting, and 2) my argument that such novel organisational forms are significantly implicated in influencing perceptions of health and illness.

Innovation as an Emergent Process

Morlacchi and Nelson's detailed analysis of the development of the left ventricular assist device (LVAD) provides a useful starting point (Morlacchi and Nelson 2011, 2016). Morlacchi and Nelson use the LVAD – an implantable electromechanic 'pumping' device used to restore blood flow as a treatment for heart failure – as a case study to delineate three components of the evolution of new medical therapies. The first of these is the generation of *new biomedical understandings of an illness*, understandings which may emerge from basic science or from clinical observation. One such event in the early history of the LVAD was the observation that the heart, like other diseased organs, can regenerate with rest. Additionally, however, medical innovation also results from an *improvement in the ability to design and use new medical technologies* – the development and introduction of suitably powerful rechargeable batteries was one such event in the history of the LVAD. This dimension of innovation often involves technology transfer: the horizontal movement of technology between industry sectors, or perhaps from one clinical specialty to another. This is of course an obvious feature of modern medicine as the origins of many now indispensable medical technologies can be traced back to developments within astronautics and aeronautics, defence, and so on. Exogenous advancements in electronics, computing, and composite materials are constantly being brought into the biomedical domain. The third aspect of medical innovation, according to Morlacchi and Nelson, is what they refer to as *learning-in-practice*. This refers to the new skills and knowledge acquired by healthcare workers as they attempt to integrate a technology or drug within an actual clinical therapy. This is not just a matter of becoming a competent user by familiarising oneself with a protocol or user manual: learning how to use a new technology entails acquiring tacit knowledge and embodied know-how that may only be learned through day-to-day clinical practice. It also refers to the collective, cumulative learning that occurs over time among groups of health professionals. An example is the ongoing refinement of surgical techniques for implanting the LVAD, and the consolidation of expertise within specialist clinical teams to support the surgeon and provide appropriate pre- and post-surgical care.

Importantly, within Morlacchi and Nelson's characterisation of medical innovation, learning-in-practice is envisaged as a source of novelty in its own right. It recognises that, in the words of Hopkins and colleagues (Hopkins et al. 2007, 568), 'the benefits of new technologies do not come from only possessing firm specific assets or competencies, but instead require the dynamic capability to effectively transform them'. New technologies must somehow be integrated within existing clinical institutional routines and clinical work 'flows'. For technologies that have considerable homologies with already in-use systems, this integration may be a relatively simple process, but for disruptive or highly novel technologies, healthcare workers will need to engage in considerable ad-hoc learning and day-to-day problem solving while accommodating various institutional pressures and professional expectations.

Morlacchi and Nelson point out that the relative influence of the three dimensions of medical innovation – new biomedical understandings, technology transfer, and learning-in-practice – will differ according to the specific therapy or procedure in question. (We can imagine that learning-in-practice is a far more influential component in the evolution of surgical procedures than it is in drug-based therapies.) They also add, however, that all three components of innovation interact strongly (Morlacchi and Nelson 2011, 523); while learning how to work with a new technology, healthcare workers often provide feedback to manufacturers on how the technology can be improved, or they may make clinical observations that advance biomedical understandings of the disease, both of which may lead to adjustments to the technology which necessitate further learning-in-practice. Morlacchi and Nelson note that the development of the LVAD, for example, entailed close linkages between cardiac surgeons and manufacturers, in which surgeons were able to provide constant feedback on the weak and strong aspects of the design of the device. It is in this fashion, according to Morlacchi and Nelson's conceptualisation, that medical practice evolves over time. Medical innovation is a cumulative, incremental process involving negotiation between various stakeholders, driven in part not only by new knowledge of disease processes, but also by technological dissemination and healthcare professionals' capacity for creativity and learning.

This aligns with the 'innovation as an emergent property' model that has been put forward by some theorists (Antonelli and Ferraris 2011; Antonelli 2009), and which, I suggest, broadly aligns with much of the work within STS. An emergent property arises from interactions within a collective or assemblage of heterogeneous agents. All agents within the collective contribute to the collective property, which cannot, therefore, be reduced to the activities or capacities of any one agent. As Morlacchi and Nelson illustrated, the evolution of useful LVAD-based therapies has been a collective process involving a network of clinicians (cardiologists, and cardiac surgeons), engineers, and industrial scientists, and the resulting therapy cannot be attributed to the activities of any one of these groups. It has, rather, emerged out of, and has been shaped by, the various interactions between them; interactions involving an onging, back-and-forth sharing of data, information, and expertise. In STS parlance we may refer to such a collective as a *network of actors*, and we would stress – following the STS analytical frame – that such networks are both mutable and temporary, as constituent actors and their various aims and interests align, change, conflict, and realign (Latour 2005, 1987).

Morlacchi and Nelson also provide a glimpse into the transformative nature of innovation. Health professionals respond to the challenges and opportunities presented by a new technology by drawing on their creative capacities and bringing in other complementary tools and protocols and expertise, thus actively transforming the terrain within which they work. Existing socio-technical networks, in other words, are reconfigured as additional allies (whether they be human or non-human) are brought in. In Nelson and Morlacchi's example, the development of the LVAD entailed the formation and consolidation of novel alliances between various professions, particularly cardiac surgeons and engineers. Other work in STS has examined the transformative effects of more 'disruptive' technologies. Pasveer's well-known study (Pasveer 1989), for example, illustrates how the introduction of X-ray imaging in medicine involved the professionalisation and institutionalisation of radiologists who, in the process, shaped how the technology was configured and how the images were used and interpreted. In a similar vien, Joyce's examination of the social life of MRI (Joyce 2006, 2008) has illustrated how a disruptive technology may be 'captured' by particular professional groups as a way of consolidating and expanding their authority within healthcare contexts. Medical innovation, then, has transformative effects on its institutional and social context, and these contexts undergo a form of constant evolution as actors respond to and utilise emerging challenges and affordances.

The capacity of actors to do this, however, is constrained by what 'innovation-as-an-emergent-property' theorists have referred to as *bounded rationality* (Antonelli and Ferraris 2011; drawing on Simon 1979). The term contrasts with the supposed Olympian rationality of actors envisaged within some economic theorisations, in which the decision-making of actors is guided by a rational appraisal of all relevant information. In reality, theorist have argued, actors within any context have only partial access to relevant information, and decision-making is informed by the limited availability of tools and resources. Actors – particularly health professionals – are subject to a host of institutional and professional pressures which constrain and shape how they respond to the emerging challenges and affordances that constitute innovation processes. Indeed, the innovation itself, whether it is a medical technology such as the LVAD or service innovation such as an electronic patient record, will consequently *reflect* such pressures; they emerge from networks of actors with constrained resources and limited capacities, and with situated, fractional (and profession-specific) understandings of the world. Innovation processes, including the innovation itself and its transformative effects, are, to some extent, structured by the existing socio-technical terrain (or what I refer to further on as socio-technical *platforms*).

Importantly, by drawing attention to the 'learning-in-practice' and 'technology transfer' aspects of the evolution of medical therapies,

Morlacchi and Nelson highlight the important role of the clinic in the innovation process. The clinic is of course the site in which clinical trials take place, surgical procedures evolve, and new service delivery and administration models are developed and trialed, but it should also be seen as a site in which technologies, whether they be drugs or devices, are dynamically and creatively transformed (rather than passively adopted) into useful clinical therapies or diagnostic procedures. In the process, latent benefits and unanticipated problems become apparent to clinicians who can thus provide feedback to other actors. The hospital, then, is not an endpoint in the process of innovation, but rather it represents a potentially powerful hub of actors within an evolving innovation network. Innovation theorists have suggested that the organisational dynamics of contemporary hospitals can be particularly facilitative of innovation. They bring various heterogeneous actors together, and provide consolidated organisational links which can permit the flow of information, expertise, and other resources. This is of course particularly the case with research hospitals with close links to academic organisations, which also have additional allocated resources for research and innovation activities (Consoli and Mina 2008). By bringing patients and patient data into controlled spaces, hospitals also provide opportunities for clinicians to make serendipitous observations, which can lead to new ideas for disease treatment or management (Djellal and Gallouj 2005). Additionally, research hospitals in particular facilitate the intergenerational diffusion of knowledge necessary for innovation – particularly tacit and non-codified knowledge – within various groups, including clinicians and non-clinicians such as managers and administrators, (Consoli and Mina 2008). (However, as we see in the next section, while such organisational dynamics certainly support some innovation activities, organisational dynamics can constrain the dynamic transformation of new technologies into useful clinical services.)

This conceptualisation of innovation-as-an-emergent-property can therefore be summarised according to several tenets. First, an innovative technology itself – whether it be a drug or a device or a novel protocol – is an emergent property of a mutable network. It is the product of ongoing interactions between heterogeneous actors, and it cannot, then, be reduced to the activities, interests, or perspectives of any one constituent actor.[3] Second, the success of an innovation depends on its capacity to appeal to, and be co-opted and moulded by, various actors who are consequently brought into an innovation network. Third, innovation is a process that entails socio-technical transformation. As actors pragmatically respond to the challenges and affordances presented by the innovation, new alliances are formed and actors themselves are reconfigured. Fourth, the capacity of actors to pragmatically respond is guided by bounded rationality: the decision-making of actors entail a balancing of various institutional and professional pressures, and it is guided by the limited access to information, tools, and resources. And, finally, the clinic itself is an important hub of innovative activities within a network. This of course not only reflects the role of the clinic as a site in which clinical trials and surgical innovation take place, but it also reflects the clinic's role as the site in which new technologies are dynamically and creatively transformed by health professionals into actual clinical services. Successful clinical adoption is about making the new technology workable within actual clinical settings, and this is itself an important part of the innovation process. The advantage of this conception of medical innovation is that it prompts us to see that clinical adoption is a collective, creative process that can result in the emergence of novel *organisational forms*; forms which reflect the perceived clinical exigencies presented by the new technology, the wider institutional constraints and affordances, and also the capacities, viewpoints, and perspectives of health professionals themselves.

For the sociologist of medical innovation, this conceptualisation of innovation-as-an-emergent-property leads us to particular avenues of inquiry. It encourages us to explore the structure of the organisational forms themselves, and the various institutional, professional, political, or social tensions, pressures, and values that are reflected in these forms. The conceptualisation also encourages us to explore the transformative effects of organisational forms. If, for example, we are to understand the way in which a novel technology is actually implicated in reconfiguring

---

[3] Although it is often the case that particular actors or groups of actors are strategically delineated as the main 'inventor' or 'innovator' – Latour has referred to this as the *secondary mechanism* of attribution (1987, 119).

understandings of health and disease, then we need to explore the organisational form which has emerged around it, and how health and disease are enacted, or rendered intelligible, within this form. We may also be encouraged to explore how this organisational form prompts changes in the wider innovation network: does it, for example, generate novel observations, protocols, or practices that appeal to, or are co-opted and modified by, other actors? It is of course these very questions that guide this book: The PMDS represents a novel organisational form that has emerged to implement DBS within therapies for severe movement disorders in children. What is particularly important about the organisational form of the PMDS is that it has been heavily influenced by the values of patient-centred care, and this, subsequently, is reflected in the way in which movement disorders – and the patients themselves – are rendered intelligible. By focusing on the PMDS, our analytical gaze is thus drawn to particular social-technical effects that we might not otherwise associate with an invasive biomedical neurotechnology such as DBS.

Further on in this chapter I introduce the notion of *proto-platforms* as a way of interrogating novel organisational forms, their co-configuring relationship with their wider innovation network, and their impact on perceptions of health and disease. However, before I do this, the following section provides a summary of some important empirical studies within the social sciences that have explored technology adoption in healthcare settings. This collection of work provides an illustration of the various forms of work that is required from healthcare professionals to 'dynamically transform' a new technology within a given setting, and it highlights some of the contextual features that constrain, shape, and facilitate this technology adoption process. More generally it provides a useful sense of the complex and highly pressurised nature of healthcare settings such as that of the PMDS.

The Problem of Technology Adoption

As mentioned in the previous section, some innovation theorists have noted that the clinic – or more specifically the contemporary hospital – is constituted by organisational dynamics that can be facilitative of innovation (Consoli and Mina 2008). Without a doubt such dynamics can be facilitative of particular innovation activities, but an emerging discourse has also framed the hospital as representing something of a problem for innovation. Within such discourse, the hospital is framed as part of a healthcare system that is insufficiently flexible to accommodate new technological developments, whether they be administration and service-related developments in IT or therapeutic developments arising from what has been called the *new biology*, such as advanced, cell-based therapies. In the UK, for example, recent initiatives to assess the emerging regenerative medicine industry have identified several features of the NHS that may hinder the 'embedding' of advanced therapies: limited capacity in existing clinical services that would support new therapies, a lack of adequate training and skills among front-line staff, and a lack of standard operational procedures for assuring quality and safety (Regenerative Medicine Expert Group 2015; House of Lords Science and Technology Committee 2013). This type of discourse tends to position the existing healthcare system as being resource-constrained and hindered by histori-cally ingrained and potentially outdated practices, and as being burdened by too much internal diversity. In the UK, technology adoption within the NHS was deemed sufficiently problematic as to warrant the establish-ment of a Department of Health-funded NHS Technology Adoption Centre (NTAC), tasked with 'understanding adoption enablers' and to work with 'selected stakeholder groups to resolve adoption barriers' (Llewellyn et al. 2014, 25).

Indeed, this problematising of technology adoption has spawned various empirical studies on adoption processes. Here, I draw on several of these empirical studies to illustrate some of the specific 'enablers' and 'barriers' of adoption within healthcare settings, and then I will briefly explore some of the theoretical propositions that have emerged from such studies, specifically the Normalisation Process Theory formulated by Carl May and colleagues (e.g. May et al. 2007).

Once a technology is 'on the market' and 'ready for use', we can envisage its adoption within a healthcare systems as involving several steps. First, the technology is somehow brought to the attention of potential healthcare providers and/or commissioners. Some healthcare systems have formal mechanisms for identifying new innovations (such as the UK's NIHR Horizon Scanning Centre), but often they will be actively championed by developers and clinician-advocates through networks with various degrees of formality. Then, at some institutional level, the technology will be subject to some sort of appraisal and a decision will be made about whether or not it should be formally adopted. In some cases, a 'test-bed', 'trial sites', or a 'demonstration project' might be commissioned to produce additional information on the technology. This appraisal and decision-making may occur within a healthcare provider (e.g. a hospital) or at a 'higher-level' governance institution, such as a Health Technology Assessment (HTA) (if it is a therapeutic or diagnostic technology) as part of an assessment of clinical and cost-effectiveness. In the former, the provider's decision will depend on their capacity to secure payment from a commissioning authority, a health insurer, or perhaps through some other fund-raising endeavour. In the latter, a positive recommendation from an HTA may led to a nationwide, publicly funded commissioning arrangement within a healthcare system – as is the case with the National Institute for Health and Care Excellence (NICE). Healthcare providers will then be expected to implement the system, although – as is the case with highly specialised therapeutic or diagnostic services – it may be that one or two providers are identified as appropriate locations for implementation.

For new technologies, procedures, and practices that we might describe as representing small, incremental adjustments to existing ones, the adoption process is likely to be relatively straightforward. The decision-making process is likely to be local, and if it does not entail a considerable cost increase over existing practices, it is unlikely that it will be subject to any formal cost-effective analysis. Its homology with existing technologies will likely mean that it can be implemented in existing work flows with minimal disruption. For many new technologies, practices, and procedures, however, the process is more complex, and a promising innovation may fail to be widely adopted within 'the clinic' because of the disruption this would have on existing organisational practices. An example of this is illustrated by Ulucanlar and colleagues' study (Ulucanlar et al. 2013) of the near-patient-testing coagulometer device, which is used to measure the blood viscosity of individuals undergoing anticoagulation therapy for various chronic conditions. Conventionally, blood viscosity is measured in hospital-based laboratory services. The device, in contrast, is portable and produces a result quickly, and can thus be used within a GP consultation. These may seem like attributes that would work in favour of its adoption, but in the UK's NHS, they brought about a level of organisational turbulence that discouraged many potential stakeholders. The adoption of the device necessitated system-wide changes, including the decommissioning of existing services, the renegotiation of contracts with community-based providers, and the establishment infrastructure for ensuring that quality and safety parameters are met (Ulucanlar et al. 2013). The device, in other words, was a 'disruptive' innovation that could not be implemented within, and adapted to, existing workflows.

Adoption within the clinic may also be hindered by differences in opinion among professional groups, and of course, when key stakeholder groups perceive the technology as being a threat to a vital aspect of their work. An example of this is the attempted introduction of telemedicine – or more specifically, the use of videophone equipment – into mental healthcare in the UK in the late 1990s (May et al. 2001). In health policy rhetoric, telemedicine was championed as a paradigm shift in the way in which healthcare was going to be organised. By enabling patients, GPs, and hospital specialists to communicate without being in physical proximity, it would 'remove distance from healthcare ... improve the quality of that care, and help deliver new and integrated services' (NHSE, cited in May et al. 2001, 1890). Psychiatry was identified as a specialty in which telemedicine would be particularly well suited: diagnostic information and treatment information could be exchanged through audio-visual communication and there is a need to extend services into under served populations (Baer et al. 1997). Yet, despite this advocacy from some groups, there was strong resistance, particularly among psychiatric nurses and occupational therapists who felt that physical proximity to patients and the nuanced interactions this enables was an important element of providing effective care. For some health professionals, then, the much-championed technology was a threat to their 'deeply embedded professional constructs about the nature and practice of the therapeutic relationship' (May et al. 2001, 1889), and the adoption of the technology became the focus of tension and conflict.

The case illustrates the importance of having front-line health professionals 'on-board' with the adoption process. Health professionals may also have varying views over what counts as sufficient evidence of an innovation's benefit to justify its adoption. Madden (2012), for example, notes that UK clinicians working within chronic wound care tend to be sceptical of the 'evidence-based' medicine being used to justify some wound care products, and they enact a distinction between this less trust-worthy evidence, and their 'real-world' clinical experience. Clinicians, she notes, are more likely to be convinced by the testimony of the colleagues (Madden 2012). Additionally, under-pressure front-line professionals seldom have the time or resources to comprehensively consult available evidence such as that provided in peer-reviewed journals (Haynes Brian 1990; Lomas 2007).

'Disruptive' new technologies, practices, and processes may also bring about financing and commissioning challenges. This is especially true for particularly novel technologies for which there may be very limited data to inform cost–impact analyses. For example, the adoption of some regulatory-approved regenerative medicines (such as ChondroCelect, a chondrocyte implantation product to treat cartilage defects in the knee) in several healthcare jurisdictions has been hindered by uncertainties over their cost-effectiveness and long-term clinical benefit: national HTA agencies have thus been reluctant to recommend them. Additionally, a demonstration of cost-effectiveness certainly does not guarantee its adoption. A relevant example here is the Breast Lymph Node Assay (BLNA), as studied by Llewellyn and colleagues (Llewellyn et al. 2014). There has been a strong case to support the implementation of this diagnostic technology within the NHS. It enables the quick 'intra-operative' analysis of breast sentinel lymph node tissue, so that if metastasis is diagnosed, auxillary lymph nodes can also be removed during the initial surgical procedure in which cancerous tissue is removed from the breast. The technology has received support from surgeons, and it also aligns with NICE guidance as it 'offers the opportunity to streamline the management of breast cancer as part of a cohesive and comprehensive service' (Yarnold 2009). The introduction of the technology was hindered by two factors. First, it had a greater disruptive impact on staff work flows within local hospitals than was originally anticipated. The consenting process for the procedure was more complex than those for existing procedures, and the added psychological strain experienced by some patients, who had woken up from their initial surgical operation to find that their cancer had metastasised, necessitated additional support from nursing staff. Implementing the BLNA, then, required additional ethical work and emotional labour. However, the major hurdle for the implementation of the BLNA was financial, and indeed the case illustrates the perverse incentives that can arise from particular payment arrangements within healthcare services. While the implementation of the BLNA would lead to cost savings for the NHS as a whole, it would also entail a loss of income for individual providers of the service (NHS Trusts), as they would be paid for only one surgical procedure instead of two (Llewellyn et al. 2014). The financial incentives of the provider, then, did not align with those of the healthcare system as a whole. The case, then, highlights that organisational payment structures may hinder the uptake of an innovation, even when the innovation has been deemed clinically beneficial and cost-effective.

It also illustrates the importance of making a good business case for technology adoption that appeals to the interests of individual providers. An attractive business case may depend in large part on the perceived prestige of the technology. This has certainly been the case for the daVinci robot which, despite its high cost and uncertainty about its clinical benefits and overall cost-effectiveness within particular surgical procedures, has been enthusiastically adopted by healthcare providers in the USA, the UK, and other countries in Europe (Abrishami et al. 2014; Tomlin et al. 2013). Its most common application is within robot-assisted radical prostatectomy, and in the Netherlands – where adoption has been high – this accounts for 70% of all prostate operations (La Chapelle et al. 2013). Adopting the device requires a considerable capital investment from healthcare providers: The purchase price is between 1.3 and 1.7 million EUR; and maintenance costs are over 100,000 EUR per year. Current evidence suggests that it does provide some clinical benefits over existing prostatectomy procedures (such as slightly reduced complications), but no clear difference in oncological results. The UK's National Institute for Health and Care Excellence, for example, has not included the device in its guidance (Tomlin et al. 2013). Yet, despite this and despite its high cost, the device provides affordances for various actors, particularly institutions. The device is perceived as highly innovative, 'state-of-the-art' and as representing the future of healthcare, and for this reason hospitals (and NHS Trusts in the UK) have been willing to invest in the technology as a way or positioning themselves as being at the cutting edge of innovative medicine (Ulucanlar et al. 2013). As Abrishami and colleagues note (Abrishami et al. 2014) in regard to the Dutch context, the robot has become a symbol for providing 'advanced care'. The robot, then, illustrates that some innovations can be endowed with meanings that have powerful performative effects on stakeholders, thus facilitating their adoption.

These various examples provide some sense of the challenges that can characterise the process of adopting disruptive innovations within healthcare settings. In his Normalisation Process Theory (NPT), Carl May has provided a more precise characterisation of the dynamics that are required to facilitate and sustain the adoption of innovations in clinical settings (May et al. 2007; May 2013b, a).[4] NPT, which has been informed by STS and in particular Actor-Network Theory, makes a series of propositions about the work that is required to embed an innovation. I won't list the propositions here, but from these, we can draw out several key points regarding the adoption of disruptive innovations in healthcare settings; points that correspond with the conception of innovation-as-an-emergent-property conceptualisation discussed earlier. First, a 'successful' innovation in medicine is not a thing in itself; rather, it represents, as May states, an organised, institutionally sanctioned, and regulated 'change in the structure and delivery of services' (May 2013a, 26). The innovative technology, practice, or procedure is actively embedded within an ensemble of other tools and practices. Second, this embedding is the result of active, ongoing work by creative agents, who utilise their own capacities and draw on available resources to routinise the technology within particular resource-constrained

---

[4] May has also produced a freely available, online NPT toolkit to assist health professionals, managers, and other stakeholder with the adoption process. See www.normalizationprocess.org/npt-toolkit/.

settings. Third, this embedding and routinising (or normalising) is a collective enterprise: diverse agents need to be *actively enrolled* in the project (i.e. the project needs to appeal to their interests), and individual workloads and goals need to be aligned with the exigencies of the innovation. An important part of this, NPT proposes, is the formation of systems of monitoring and evaluation: it permits collective, reflexive appreciation and ongoing re-enrolment of health professionals. And, fourth, the embedding and routinisation is a transformative process. As agents are re-orientated and aligned into new collectives and as new ensembles are created, and as actors explicitly differentiate the innovation as being superior to pre-existing arrangements, the very terrain that constitutes healthcare is of course altered. New organisational forms emerge that reflect the perceived exigencies of the innovation itself, the institutional contexts within which it is routinised, and the capacities and perspectives of the collective of agents involved in its routinisation.

In subsequent chapters we see how the PMDS team has creatively managed some of the challenges that have been touched upon in the examples provided earlier, and how their unique organisational form has emerged as a result. What is particularly interesting with the PMDS is that this organisational form reflects many of the values conventionally associated with patient-centred medicine. In the process of managing some of the challenges of adoption, and in the process of taking advantage of the affordances presented by DBS, the team has creatively drawn on their capacities and pulled together tools and resources from professions – such as occupational therapy – that lay a heavy emphasis on patient-centred care. This has taken place in a hospital which was specifically designed to encourage comprehensive care for families, and within an organisational payment structure that permits this. As a consequence, DBS has become embedded within patient-centred clinical service.

This, I demonstrate, is reflected in the transformative effects of implementing DBS in paediatric neurology. It is reflected in the way in which the disease and the patient are rendered intelligible within the service, and it is reflected in the way in which this local organisational form is itself prompting changes in the wider context of paediatric neuromodulation. In the next section of this chapter, I provide a conceptual frame for making sense of these transformative effects. I suggest that the organisational form of the PMDS can be understood as an example of what I will define as an emerging patient-centred *proto-platform*. By bringing together Keating and Cambrosio's notion of *biomedical platforms* (Keating and Cambrosio 2003), with Schot and Geels' (Schot and Geels 2007) notion of *innovation niches*, it provides a useful analytical frame for making sense of how innovation at the point of technology adoption is both structured by, and has structuring effects on, the surrounding context.

Towards a Conceptualisation of Proto-Platforms

The previous section provided illustrations of some of the specific challenges entailed in technology adoption. From this we get the picture that health-care professionals are creative, pragmatic, and strategic agents, but that they are anchored in resource-constrained contexts which – to use the policy phrasing – presents various 'enablers' and 'barriers' for their technology-adoption activities. However, to state it in this way, that a context presents various 'enablers' and 'barriers' for agents, does not capture the degree to which contextual elements, whether they be material (architectural spaces, other technologies, and objects) or 'cultural' (institutional strategic aims, commissioning arrangements), can shape their activities, and can thus have a structuring effect on the emergent outcome. This is not to say that such elements determine the actions of actors. Rather, it is to say that contextual elements themselves should be seen as active agents in the technology adoption process: they constrain, shape, channel, prompt, and stimulate human agents in ways that are often unanticipated, and that cannot necessarily be reduced to the interests of other immediate human agents. This aligns with the 'innovation-as-an-emergent-property' conceptualisa-tion outlined earlier, and it is of course a key tenet of STS: that human agents are both culturally and materially situated (Mol 1999; Law 2009), and that a comprehensive, sociological explanation of action cannot not appeal to 'paradigms', 'ideologies', and 'ideas' alone (Latour 2005): it must also account for the various non-human actors, or what Latour has called 'the missing masses' (Latour 1992).

This structuring effect of various contextual elements – both material and cultural – on innovation activities in medicine has been explored in

Keating and Cambrosio's *Biomedical Platforms* (Keating and Cambrosio 2003). Keating and Cambrosio coined the notion *biomedical platforms* to describe the emergence of a particular infrastructure and associated set of practices for diagnosing blood cancers; *biomedical* infrastructure and practices which have come about via the merging of developments in biology and medicine over the past 60 years and which represent, they argue, a significant transformation in the way in which the normal and the pathological are rendered intelligible. The notion is particularly useful for my endeavour in this book. It draws our analytical attention to the distinctive role of non-human agents in structuring innovation and in reconfiguring understandings of disease and illness while avoiding the reductionist pitfalls of either technological determinism or cultural determinism. And, in doing so, it also provides a means for under-standing the articulation between routine clinical work and innovation.

Specifically, Keating and Cambrosio trace the development of immu-nophenotyping and its adoption into routine haematological services (or, more precisely, they trace the emergence of haematological cancer services centred on the routinisation of immunophenotyping). Within these services, blood cancers are diagnosed using practices which involve the delineation of blood cell types according to particular, standardised immunological markers within cell membranes. Using these markers to characterise cell types, abnormal blood cell morphologies are identified and populations of blood cell types can be separated, counted, and compared to established norms. This reliance on particular biological entities to make sense of disease, Keating and Cambrosio note, is indicative of contemporary biomedical practice more generally, and it also stands in contrast to medical practice in the past when physicians tended to prioritise the narratives and biographies of their patients (Pickstone has referred to this as biographical medicine (2001)). Importantly, they note, this type of medical work requires a vast array of tools, technologies, and standards, particular types of spaces and architectural forms, and of course, ways of thinking. For example, the emergence of these haematological services has entailed construction of particular architectural arrangements, in which laboratories are situated in close proximity to patient wards within hospitals, and various systems to enable the flow of biological samples and information between them.

It has entailed the establishment of international standards for diagnosing cancers according to immunophenotypes, and the creation and dissemination of immonphenotyping equipment, such as the flow cytometer. It has also entailed the emergence of clinicians and laboratory technicians with particular skill sets and expertise. And it has entailed the development and refinement of regulated routines and protocols which bring these non-human and human actors (patients, clinicians, laboratory technicians) together within particular configurations. It is these various elements and practices that constitute what Keating and Cambrosio define as biomedical platforms. In haematological services, they can be defined as:

> specific combinations of techniques, instruments, reagents, skills, constituent entities (morphologies, cell surface markers, genes, spaces of representations, diagnostic, prognostic and therapeutic indications, and related etiologic accounts). More specifically and more abstractly [they are] material and discursive arrangements that act as a bench upon the conventions concerning the biological or normal are connected with conventions concerning the medical or the pathological. (Keating and Cambrosio 2003, 4)

Platforms, then, can be thought of as a socio-material infrastructure, and as a particular way of doing things and making sense of the world that is embedded within this infrastructure. Collectively, the elements that constitute a platform participate in rendering disease and illness intelligible in a particular way; as, for example, a markedly raised level of white blood cells that characterises some leukaemias. We could say, then, that such biomedical platforms configure a particular type of biomedical gaze – a biomolecular gaze (Bell 2013) in the case of blood cancer platforms – by permitting and facilitating knowledge-producing practices that foreground particular biomedical attributes, while numerous other phenomena, such as a patient's biography, are elided or ignored.

Keating and Cambrosio illustrate that such biomedical platforms provide a structuring scaffold for subsequent innovation activities. The immonophenotyping platform which had originally emerged to diagnosis blood cancers has been deployed by researchers to make sense of HIV/AIDS. From the onset of the epidemic, the embedded platform provided a ready, near-at-hand set of resources and 'ways of understanding' that researchers and clinicians could tap into, and this is reflected in how AIDs is now diagnosed and how HIV is understood: the 'T-Cell count' measure that has become the common measure of 'health' for afflicted individuals is an example of this (Keating and Cambrosio 2003, 8, 269–275). Drawing on some of the terms introduced earlier, we can say that a new biomedical platform – an HIV diagnostic platform – has emerged from a process of technology transfer and learning-in-practice, as researchers and clinicians have pragmatically drawn on various elements of pre-existing platforms. This new platform is consequently closely entwined with that from which it is emerged. As Keating and Cambrosio put it:

> New platforms are articulated and aligned in complex ways with existing ones, thus integrated into an expanding set of clinical-biological strategies. (Keating and Cambrosio 2003, 4)

Platforms, then, represent embedded, routine clinical practice and a structuring springboard for subsequent innovation.

In this book, I adopt this concept of platforms to explore the organisational form that has emerged to deliver DBS in paediatric neurology. The PMDS can be seen as representing a platform that has evolved from the clinical adoption activities of health professionals as they pragmatically and creatively draw on elements of their existing socio-material context. And collectively, the various elements of this emerging platform participate in rendering disease and illness in particular ways; they foreground some elements of the patient, while eliding and ignoring others. The book, then, examines various socio-material elements of this emerging platform and the way in which they configure understandings of health and illness. It examines some of the platform elements explored by Keating and Cambrosio: architectural forms and spaces for clinical work; infrastructures for diagnosing and assessing illness; and 'ways of understanding'. It also explores platform elements that are not explored in detail in Keating and Cambrosio's work, particularly the embodied knowledge of clinicians, and tools for aligning visions of the future. What is particularly interesting about the PMDS, however, is that many of these elements reflect what are often championed as being 'patient-centred' values, and this is reflected in what I have defined as the broad clinical gaze that is enacted within the PMDS. The clinical adoption of DBS has, in other words, brought about what could be called a patient-centred platform, and this platform has some interesting and potentially unexpected social effects.

The biomedical platforms explored by Keating and Cambrosio have become relatively widespread in contemporary medical contexts: immunophenotyping platforms are characteristic of specialised hospitals around the world. Indeed, an important characteristic of platforms is that they traverse individual clinical sites: they involve infrastructures of standardised entities, regulations, and formalised protocols that enable coordination across geographical space. As a pioneering service, the PMDS, in contrast, is one of very few centres worldwide that specialise in providing DBS in a paediatric context. The PMDS organisational form is localised and relatively unique, and indeed one of the challenges for the PMDS is to establish a set of infrastructures and alliances that traverse paediatric neurology so that their accomplishments can be recognised and DBS can subsequently be more widely adopted. For this reason, I refer to the PMDS as representing a *proto*-platform: it is a nascent platform that is to some extent precarious, but which, via specific activities of PMDS team members, is becoming increasingly consolidated and is prompting some wider changes in paediatric neurology.

It is here that Schot and Geel's notion (Schot and Geels 2007) of *innovation niche* is useful. It provides a means of conceptualising the movement from proto-platform to a more consolidated platform such as those explored by Keating and Combrosio, and the transformative effects of this movement. Working within a broadly 'innovation-as-emergent-property' understanding, Schot and Geels use the notion of innovation niche to articulate how the adoption of disruptive technologies can bring about social change. Some innovations, they note, are perceived by stakeholders to be so disruptive that they are, to some extent, incommensurable with existing infrastructure, or what they refer to as *socio-technical regimes*. For these innovations to be successful, it is necessary for stakeholders to carve out a protected space – an innovation niche – in which the innovation can initially be implemented and nurtured so that additional resources can be mobilised. The niche is collectively constructed, and it represents the socio-technical supporting infrastructure, skill sets, and 'ways of thinking' that are necessary for its implementation, and which may not easily integrate with existing infrastructures and work flows. Depending on the ability of the innovation to increasingly appeal to the aims and interests of a wider array of actors, it may be that the niche itself expands as more and more actors are enrolled, to the point where it eventually becomes a new socio-technical regime, perhaps supplanting previous regimes. In healthcare, this may entail the establishment of new regulatory frameworks and reimbursement structures, as well as new clinical services and health professional alliances (see Gardner and Webster 2016). This is not a process of one-way imposition, however: elements of an expanding platform will themselves be adapted to some degree as heterogeneous actors pragmatically and creatively mould them to align with their own interests. It is in this way, according to Schot and Geels (2007), that innovations become widespread and routinised, and it is in this way that innovations are transformative, as well as being – to varying degrees – adapted. Hence, the diffusion of disruptive innovations in medicine entails the concurrent emergence of new socio-technical networks for operationalising, managing, and evaluating them.

A *proto*-platform can thus broadly be understood as an innovation niche. It has emerged to support a disruptive innovation and it is driven by promissory expectations. It may or may not, depending on the ability of stakeholders to rally additional actors, become a platform that traverses 'pioneering' local sites. In this book I conceptualise the proto-platform of the PMDS as a carefully crafted space in which DBS can be initially implemented and refined. The success of the technique ultimately depends on the further expansion of this space. It requires, in other words, the dissemination of supporting proto-platform elements: supporting technologies, ways of thinking, embodied knowledge, and so on, and these elements will, in the process of traversing local sites, undergo some degree of adaption and adjustment. In light of this, this book will also attend to the transformative potential of the PMDS organisational form: the team is highly regarded among the wider paediatric neurology community, and some of their activities are aimed at prompting alliances to produce the evidence that could support the dissemination of DBS. These include, for example, an attempt to popularise particular patient-centred clinical assessment tools and thus create a shared understanding of what counts as 'suitable evidence'. If successful, such practices would help bring about the establishment of a more widespread platform. This would, of course, also entail a further embedding of the broad clinical gaze.

Creativity, Constraint, Uncertainty

We began this chapter with a critical discussion of the linear model of medical innovation. A significant deficiency of the linear model, I argued, is that it elides the complex, creative, and indeed innovative work that is entailed in adopting new technologies within clinical contexts. In light of this, I provided an alternative conceptualisation of innovation: the 'innovation-as-an-emergent-property' conceptualisation. This is a thoroughly sociological understanding of innovation that aligns with key tenets in STS. It draws the analytical gaze to the socio-material networks of actors (possessing a bounded rationality) that constitute innovation activities, and to the interweaving of various processes that bring about an evolution in medical practices; processes that Morlacchi and Nelson (Morlacchi and Nelson 2011, 2016) have usefully conceptualised as the generation of new biomedical knowledge, learning-in-practice, and improvements in the ability to design and use new medical technologies (which includes technology transfer). Most importantly, this innovation-as-an-emergent-process conceptualisation foregrounds the important role of 'the clinic' as a site of innovation activities, not just as the site of clinical trials and surgical innovation, but also as the site at which new technologies, practices, and procedures are dynamically transformed by creative agents into workable and working services.

We then narrowed our focus to this 'problem' of technology adoption in healthcare settings. We surveyed some of the empirical studies that have been conducted by social scientists and, in doing so, I highlighted the types of challenges that can characterise technology adoption processes. These relate to the difficulty of integrating an innovation within the workflows and organisational routines of local sites, professional interests, and a lack of shared understandings among health professionals (and other stakeholders), and financing and commissioning challenges, such as perverse financial incentives for healthcare providers and uncertainties around cost-effectiveness. These are challenges and tensions that characterise the context of the PMDS, and in the following chapters, we will explore some of their attempts to manage these. We will see that PMDS members have engaged in the type of work that, according to May's Normalisation Process Theory (May 2013a), is required to embed a disruptive innovation such as DBS within a clinical setting. In Chapter 4, for example, we explore some of the administrative practices of the team that provide a space for communal appraisal of DBS therapy, and that align individual and team work schedules with the exigencies of DBS.

I have also suggested that the PMDS organisational form that has emerged as a result of such embedding activities can be conceptualised as a manifestation of a proto-platform: it is a nascent, somewhat precarious but potentially consolidating and expanding example of a platform as conceptualised by Keating and Cambrosio (2003). Using this as an analytical frame, in the following chapters we explore some of the specific socio-technical elements that constitute the proto-platform. In Chapter 4, in addition to exploring some of the administrative practices of the team, we briefly explore the architecture of the hospital within which the PMDS team is based. In Chapters 5 and 6, we explore important platform features that have not received much attention in Keating and Cambrosio's account: the embodied knowledge (particularly what I refer to as communicative body work and sensorial reflexivity) of some of the PMDS team members (Chapter 5); and tools and practices for aligning visions of the future (Chapter 6). In Chapter 7, we explore PMDS tools and practices for assessing the clinical effectiveness of DBS. We also see how the team is attempting to create a broader consensus about the effectiveness of DBS by popularising these practices and tools among the wider paediatric neurology community. Each of these elements, we shall see, have emerged as PMDS team members have attempted to manage specific clinical problems.[5]

---

[5] A brief overview of methodology and data collection methods is described in 'Notes' on page 215.

And I have suggested in this chapter that this proto-platform of the PMDS enacts disease and illness in a particular way. In each of the chapters we will see how the various elements in the platform participate in the perpetuation of what I have defined as a broad clinical gaze. This, as we will see, has some very interesting implications for the way in which patients and their illness are rendered intelligible. The therapy provided by the PMDS involves putting electrodes into the brains of children, and they rely heavily on technologies (such as MRI) that are known to perpetuate neurocentric notions of personhood and illness (Joyce 2008). Yet, in many of their interactions, the clinical gaze of the PMDS enacts a broader conception of personhood and illness, or what could be called a biopsychosocial conception of personhood and illness. As we will see, this is in large part due to the way in which many of the platform elements reflect ideals and values that are often associated with patient-centred medicine.

The PMDS, then, provides not only a very good case study for exploring clinical adoption, it also provides a useful case study for exploring what patient-centred medicine actually looks like in a specific context, and for exploring the cultural implications of an important new neuro-technology. Indeed, as I stated at the outset of this chapter, a premise of this book is that if we want to comprehensively explore the cultural impact of new technologies, then we need to examine very closely how, in the process of being creatively and pragmatically implemented within actual clinical services, they become embedded within wider socio-technical systems. On this point, I follow the same vein as Jessica Mesman's study (Mesman 2008) of medical innovation in neonatal care. Mesman frames her book as an act of *exnovation*. This, she states, is a process of foregrounding what is already present in specific practices. It means:

> to bring to light implicit matters or actual practice and to develop a fresh perspective on the ingenuity of the professionals and the specific structure of their practices. (Mesman 2008, 5)

Similarly, this book can also be seen as an act of exnovation. It brings to light the specific, 'routine innovation' practices of specific health professionals that tend to be elided by some prevalent conceptions of innovation. By doing this, as Mesman argues, we highlight the limited power of medical technologies: their transformative effects – whether they be desirable or otherwise – depend on how they are dynamically, actively worked into the world by various agents.

Mesman also emphasises a key dimension of this 'routine innovation' work that has received little attention in the proceeding discussion, but which we need to keep in mind as we explore the PMDS. She notes that innovation, particularly the work that is required in adopting and managing new, high-tech biomedical innovations, occurs in clinical settings that are plagued with uncertainty. My account of the PMDS that follows is one in which the PMDS, at least at this initial stage, has been relatively successful in adopting DBS technology, and my illustrations of PMDS teamwork, particularly in Chapters 4 to 7, focus predominately on relatively straightforward clinical cases. Because of this the sense of uncertainty that might have characterised the PMDS is somewhat lost. It is important to remember that afflicted bodies and patients, their families, and the technologies themselves, can be delicate and erratic, and collectively they constitute an environment that will always escape the mastery of health professionals. This, compounded with their own fallibility and the pressures of balancing institutional and professional commitments, means that we should see health professionals as muddling through, day to day, as best they can.

3

A History of Deep Brain Stimulation

In light of the discussion of innovation in the previous chapter, we should not see the PMDS proto-platform as an inevitable product of scientific progress, nor as the consequence of 'profound intentions and immutable necessities', to borrow Foucault's phrasing (cited in Rabinow 1991, 89). Rather, we should see the emergence of the PMDS as being contingent upon a conjunction of circumstances involving a range of actors with diverse interests. We should, in other words, see the PMDS as being depended upon particular historical *conditions of possibility*. While the last chapter examined ways in which innovation can be conceptualised, this chapter explores the actual innovation processes that brought about these *conditions of possibility*. This chapter, to put this more concisely, is an account of the history of DBS, and thus provides useful background to the activities of PMDS.[1] This enables us to better comprehend the tensions that characterise the day-to-day work of the PMDS, and the significance of their patient-centred approach.

---

[1] A description of the source material for this historical account is provided as appended in 'Notes' on page 217.

© The Author(s) 2017
J. Gardner, *Rethinking the Clinical Gaze*, Health, Technology
and Society, DOI 10.1007/978-3-319-53270-7_3

We specifically focus on the history of the DBS technique, paying particular attention to the various contingent circumstances that underlie its development and dissemination. There are of course quite a few overviews of the emergence of DBS within the scientific literature (Sironi 2011; Danish and Baltuch 2007; Talan 2009), but in most cases, these are seldom more than a means of introducing the technology to readers, and they tend to provide an account that would align with the overly simplistic linear model of innovation. As such, they tend to be brief, and they tend to rely on a few key historical 'discoveries' to provide a sense of context. As Hariz and colleagues (Hariz et al. 2010) have noted, the common narrative of these overviews is that DBS was first developed in 1987 by a team treating patients with tremor at Grenoble, France. Alim-Louis Benabid and his colleagues were using electrostimulation to identify the correct area of the brain to undergo ablative surgery. In doing this, they noted that high-frequency electrostimulation was, on its own, therapeutic, and subsequently, DBS was adopted as a means for treating medically refractory motor disorders in many parts of the world. And as time went on, the narrative goes, the mood-related side effects identified in some patients encouraged explorations of DBS as a possible treatment for psychiatric illnesses. While the details of such narratives may not necessarily be wrong, they tend to give a misleading impression of the nature of medical and scientific innovation. Much of the scientific literature presents DBS as if it is an inevitable outcome, the consequence of continuous development driven by good science, beneficent medicine, and committed specialists.

DBS, however, has a complex history. As we will see, the DBS technique was first developed as a treatment for chronic pain: an indication for which it currently does not have regulatory approval. Adhering to the 'innovation-as-an-emergent-property' conceptualisation, this chapter will explore this complexity, drawing attention to the conjunction of circumstances that have shaped its development trajectory and brought us to that point in time when the PMDS came into being. In the process, we see that the emergence of DBS has been heavily dependent on 'technology transfer' and 'learning-in-practice'. It has been propelled by the charisma and creativity of particular clinicians, strategic clinician–industry alliances and interestingly, serendipitous accidents. We also see that the development and dissemination of DBS in its current form has been contingent upon concurrent and interweaving developments in pharmaceuticals, imaging technologies, medical device regulation, and most importantly, I argue, standardised clinical assessment tools; tools which have enabled the extraction of and pooling together of quantitative data from heterogeneous bodies.

Throughout this chapter, then, I draw out various themes that were touched upon on the last chapter. Some of these are particularly relevant for the exploration of the PMDS activities in the following chapters. The first of these is that a range of actors has been willing to invest their time and resources into DBS because it has appealed to professional and institutional interests. It has, in other words, strategic utility, and actors have been able and willing to coordinate their own routines around it. Second, the establishment of shared understandings about what constitutes 'evidence' of DBS clinical effectiveness has been vital for its dissemination. It has enabled the communal appraisal of the technique, and it has enrolled other actors into the project. Later in this book (Chapter 7) we see that the PMDS is working within these pre-existing shared understandings, while also attempting to establish new shared understandings about 'what counts as evidence'; evidence that captures clinical changes, PMDS team members argue, that patients themselves feel are important.

I have structured the following historical account into sections that are roughly chronological, each of which focuses on an important aspect in the history of DBS. The first ('The Rise and Near-fall of Stereotactic Neurosurgery') describes the emergence of stereotactic neurosurgery that precipitated early neurostimulation techniques. In the second section ('The Birth of the Neurostimulator') I outline the development of the neuro-stimulator, drawing attention to its relationship with the cardiac pacemaker industry. I then outline the advent of medical device regulation and the impact this had on the then quickly disseminating neurostimulator. As I outline in the section 'Clinical Trials, Evidence, and Clinical Assessment Tools', this new regulatory climate necessitated new technologies for measuring clinical effectiveness. The development of these technologies in particular disease areas determined the subsequent application of the technique, as illustrated by the work of Alim-Louis Benebid's clinical team. We also examine the FDA's decision to grant marketing approval to specific applications of DBS. In the final section, we look more directly at the implications of these historical developments on the activities of the PMDS team, including the problematic reimbursement climate for DBS in dystonia.[2]

The Rise and Near-Fall of Stereotactic Neurosurgery

As stated earlier, much of the existing literature on the history of DBS gives the impression that DBS has its origins in the 1987 work of Alim-Louis Benabid's team, based in Grenoble, France. Yet, neurostimulators had been under development since the 1960s (the term 'deep brain stimulation' was first trademarked by Medtronic in 1975), and 'modern' clinicians had been aware of the therapeutic effects of neurostimulation for several decades prior to this. There are also anecdotal reports that electrostimulation was used in ancient and medieval times: Roman Emperor Claudius, for example, suggests in his *Compositiones Medicamentorum* that a severe headache could be alleviated by applying an electric ray directly to the cranium, and there are various reports that other electric fish were used in an attempt to treat various ailments until the eighteenth century (Sironi 2011; Rossi 2003).

It is in the 1930s that what we might call the modern era of neurostimulation gained traction. Since the 1930s, clinicians had been using neurostimulation as an exploratory probe within ablative surgeries; i.e. surgical procedures in which specific parts of the brain are deliberately destroyed in an attempt to treat neurological or psychiatric dysfunction. The most well-known of these procedures during this era is Wilder Penfield's 'Montreal Procedure', developed as a treatment for epilepsy. During the procedure, patients are kept awake while the surgeon attempted to identify which part of the cerebral cortex was malfunctioning using an electric probe. The probe would be used to stimulate a discrete area, and depending on the patient's response, the surgeon would decide

---

[2] Much of the following content of this chapter has been published in an earlier form: Gardner, J. (Gardner 2013), A history of deep brain stimulation: Technology innovation & the role of clinical assessment tools. *Social Studies of Science.* 43(5): 707–728.

whether or not that area required ablation (Penfield 1936). The procedure became well-known, and in Montreal over 2000 patients were treated in this way. It was apparently effective in over half of those treated. In the following decade, a technology was developed that would allow this electrical probing and ablation technique to be conducted on areas deeper within the brain. This was the stereotactic apparatus, and its genesis brought about a neurosurgical subspecialty – stereotactic neurosurgery – which is still practiced today, and it is within this subspecialty that many of the skills, knowledge, and equipment were first developed that would later inform the development of DBS.

In effect, the stereotactic apparatus enabled surgeons to map and navigate the brain. It does this by delineating the brain as a three-dimensional Cartesian matrix. Using available brain imaging techniques, the three-dimensional system enabled any point of the central nervous system to be designated as a set of coordinates.[3] The surgeon could carefully map the position of sensitive brain structures and navigate their way to areas deep within, such as the basal ganglia, and ablate specific parts (Spiegel et al. 1947). With the introduction of the apparatus, the mortality rate associated with neurosurgery plummeted from 15% to 1%, and the stereotactic subspecialty went through a period of rapid growth. Within 10 years of its introduction, it was being practiced in over 40 centres worldwide (Gildenberg 2000).

The apparatus was used for tumour removal, but a significant reason for its rapid uptake was the considerable demand for ablative therapies. During the late 1940s and early 1950s, it was the only treatment available for a range of crippling neurological and psychiatric conditions, including chronic pain, movement disorders, and schizophrenia. Prior to the introduction of chlorpromazine (the first antipsychotic medication) in the 1950s, ablation and the much cruder (and even at that time, much criticised) frontal lobotomy were the only available treatments for patients with what could be crippling psychiatric disorders. As Mashour and colleagues (Mashour et al. 2005) note, these disorders were considered

---

[3] Up until the 1970s, brain images were produced using pneumoencephalography; an uncomfortable procedure in which cerebral spinal fluid is drained from around the brain, so that it would produce contrasts that could be visually rendered via X-ray.

to be a significant financial burden on society, particularly in the USA where, in the early 1940s, over half the hospital beds were occupied by psychiatric patients. The situation was similar to movement disorders such as Parkinson's – a disorder that affects three people in every thousand. Pharmaceutical-based treatments for Parkinson' were not introduced until the mid-1960s, meaning that there were tens of thousands of patients with debilitating symptoms that were otherwise untreatable. Stereotactic-guided ablative surgeries were therefore quickly adopted. Surgeons would roughly follow the probing technique devised by Penfield: the stereotactic apparatus would be used to guide the surgeon to the parts of the brain that were thought to be malfunctioning; an electrical probe would be used to mimic the ablation and test the response; and a permanent ablation would then be made. By the time that levodopa pharmaceutical treatments for Parkinson's were introduced, over 25,000 patients worldwide had received ablative therapy (Gildenberg 2000).

Hence, for approximately two decades, there was a large reservoir of what were otherwise untreatable patients. This provided stereotactic neurosurgeons with a large amount of work and with considerable opportunity to engage in what we referred to in the last chapter as 'learning-in-practice' (Morlacchi and Nelson 2011). During the ablative procedures and while hunting for the most efficacious target, surgeons were able to explore the effects of electrostimulating different areas of the brain, including the basal ganglia. This is nicely articulated by Gildenberg, who describes it as:

a period of unrivalled empirical human experimentation . . . From the beginning, the philosophy was to use every insertion of an electrode into the brain as an opportunity to study neurophysiology . . . The information obtained in the operating room was valuable to help localise the electrode position, and the information obtained about pathophysiology was used to develop new indications and targets for stereotactic surgery (Gildenberg 2000, 299, 301).

As a result, surgeons were able to build up a repository of knowledge about neurostimulation and its effects on particular regions of the brain. For example, it was noted that low-frequency stimulation of some areas such as the subthalamic nucleus (STN) – an area within the basal ganglia – could exacerbate motor disorder symptoms, while higher-frequency stimulation could reduce symptoms (Mundinger 1965; Nashold and Slaughter 1969). Towards the end of the twentieth century, surgeons would later revisit much of this work – indeed, as we shall see further on, the STN was subsequently confirmed as the most efficacious target for managing Parkinson's tremor.

By the end of the 1960s, however, this era of 'empirical human experimentation', and indeed the stereotactic neurosurgery subspecialty, was almost brought to an end. Clinical and political developments drastically reduced the pool of patients. The success of levodopa pharmaceutical treatments for Parkinson's meant that neurologists were reluctant to refer patients to neurosurgeons for what were highly invasive, risky procedures. Levodopa appeared to be safe, it was relatively affordable, and it appeared to be remarkably effective in managing Parkinson's systems. For these reasons it very quickly became the first line of treatment. Similarly, chlorpromazine was being used with some success to manage many patients with psychiatric disorders. Many patients did not respond favourably to antipsychotics, but neurosurgeons in the USA were discouraged from treating psychiatric disorders by an influential public campaign led by the psychiatrist Peter Heath. Heath made no distinction between stereotactic-assisted psychosurgery and the deeply problematic frontal lobotomies like those conducted by Walter Freeman, claiming that they have 'no empirical or rational basis ...' and that they 'attack and mutilate brain tissue that has nothing demonstrably wrong with it'. He also told a US Senate Subcommittee that such procedures were being used to 'subject the individual to the control of others' (Breggin 1972, 381). A commission of inquiry was launched in response to the public campaign, and although it concluded that psychosurgery appeared to benefit many patients, the political climate and prevalent public opinion effectively led surgeons to abandon the field (Valenstein 1997).

For these reasons, by the mid-1970s stereotactic neurosurgery had experienced a significant decline. Worldwide, a few specialist academic centres continued to practice stereotactic surgery to provide ablative therapy to the small number of patients with movement disorders or chronic pain that failed to respond satisfactorily to pharmaceutical-based treatments. It was in these centres that the skills, knowledge, and technology that had been developed during the earlier era of prolific experimentation were maintained, and indeed it was within these centres that the DBS technique as we know it today would eventually have its genesis.

The Birth of the Neurostimulator

While it has become apparent to surgeons that electrostimulation of certain brain areas at certain frequencies has therapeutic effects, neurostimulation had not become a therapy in its own right. The obvious reason for this is that the technology available at the time did not permit it: electrodes were large and bulky, and it was necessary for them to protrude from the patient's skull so that they could be connected to a power source. These too were large and bulky, and certainly not implantable. As noted in the previous chapter, important advancements in medical practice often result from technology transfer – the horizontal movement of technology from one specialty to another (Morlacchi and Nelson 2011). The neurostimulators, which had its genesis in the emerging, lucrative cardiac pacemaker sector, is an obvious example of this.

The first commercially available pacemaker was introduced in the early 1960s by the fledgling medical device manufacturer Medtronic. It proved to be a significant clinical advancement, and for Medtronic, which subsequently became one of the world's major medical device manufacturers, it was a hugely lucrative commercial success. It was from this activity that that the neurostimulator arose: its development was fuelled and shaped by the technology, skills, and indeed financial resources of the cardiac pacemaker industry. The transferability of the cardiac pacemaker was first illustrated in the late 1960s. The Wisconsin-based neurosurgeon Norman Shealy adapted a pacemaker so that it could be used to stimulate the spinal cord as a treatment for chronic pain. Initially, a small group of patients had electrodes attached to their spinal cord, which were then connected to an implantable Medtronic pacemaker, modified so that it would produce a higher-than-normal frequency of stimulation (Shealy et al. 1967). The pacemaker was connected to an externally worn power source via a transcutaneous radiofrequency system. The results of the permanent stimulation, Shealy reported, were mostly promising, and the method was soon taken up at other clinical centres in the USA (Shatin et al. 1986). Not long after, a Californian team with expertise in stereotactic surgery used modified pacemakers to treat chronic pain by stimulating areas deep within the brain (Hosobuchi et al. 1973).

By the late 1960s the rising interest among clinicians encouraged the medical device industry to develop specific neurostimulator devices. The first to do so was Medtronic in 1968, followed soon after by Avery Laboratories and Cordis. In effect, these companies utilised and adapted their existing cardiac pacemaker manufacturing platforms to produce neurostimulators: the devices, then, were similar in appearance and shared many basic technical features. Additionally, these companies subsidised the production of neurostimulators with the huge financial returns from the cardiac pacemaker (Upton 1986). In a significant move, in 1975, Medtronic established its neurological division. It was at this time that it trademarked the term 'Deep Brain Stimulation' (Coffey 2009).

In the 1970s, these early neurostimumators were used in a number of specialised centres in Europe and the USA. In addition to chronic pain management, surgeons incorporated the neurostimulator in treatments for a range of conditions that had previously been shown to respond to ablative therapy: cerebral palsy and movement disorders such as essential tremor, Parkinson's, and dystonia, epilepsy, as well as psychiatric conditions, particularly severe depression and schizophrenia. The specific target of stimulation varied according to the condition and current knowledge. For some treatments it was the spinal cord or the cerebral cortex, but in those specialist centres that had retained stereotactic capability, areas deep within the brain such as thalamus were also stimulated. The work of the charismatic neurosurgeon Irving Cooper is a useful illustration here. He used neurostimulators to stimulate the cerebellum and in some cases areas within the basal ganglia to treat dystonia, epilepsy, and cerebral palsy (Cooper and Upton 1978; Cooper et al. 1982, 1980). Approximately 200 patients received these treatments, and Cooper proclaimed that the results were good and 'worthy of immediate notice' (Rosenow et al. 2002). Similarly in Germany, Fritz Mundinger used Medtronic neurostimulators to stimulate areas within the thalamus to treat dystonia, arguing that the reversible nature of the treatment made it preferable to ablation (Mundinger 1977). This statement was echoed by Orlando Andy of Mississippi who was using stimulation of areas within the thalamus to treat nine patients with Parkinson's who had failed to respond to levodopa therapy (Andy 1983). At Tulane University, Robert Heath adopted Medtronic neurostimulators into his treatments for psychiatric disorders, particularly schizophrenia, and reported that some of these patients subsequently became symptom free (Heath 1977; Heath et al. 1980). In Southampton UK during the late 1970s, Brice and McLellan were using DBS to treat a small number of patients with multiple sclerosis–associated intention tremor. They reported that some of their patients improved significantly: one patient who was severely disabled was subsequently able to 'feed herself, light her own cigarettes, fasten her own buttons, and control bed light and radio' (Brice and McLellan 1980).

In his retrospective, Philip Gildenberg gives us some idea as to why neurosurgeons were receptive to neurostimulation technology. The introduction of levodopa for the treatment of Parkinson's in 1967 had left a work void for functional neurosurgeons with training in stereotactic techniques. Neurostimulation provided an opportunity, for some neurosurgeons at least, to utilise their skills and equipment and provide potentially effective surgical treatments for other conditions or for those few patients with Parkinson's who could not tolerate levodopa (Gildenberg 2009, 14). The skills, knowledge, and tools that had been developed during stereotactic surgery's earlier period of rapid growth could easily be transferred into DBS therapies: the stereotactic apparatus in conjunction with imaging technologies were used to identify target areas and plan the surgical procedure to implant permanent electrodes; intra-operative stimulation was often used to ensure that the correct target area had been located; and these target areas were often the same as those that, in the past, would have been ablated. Neurosurgeons, then, had the necessary skill set, and the technology provided them with a vehicle for intervening in complex neurological conditions in an era when drug-based therapies tended to dominate.

This dissemination was enabled by the flexibility of the technology. As Faulkner argues, the material qualities of a device interact in more or less flexible ways with social actors, impinging upon their possibilities for adoption and usage (Faulkner 2009, 18). The material qualities of neuro-stimulation technology, such as its small size, its biocompatibility, and its ability to deliver precise electrical stimulation to a region decided upon by the clinician, permitted its adoption into a range of therapies. This is not to say that the diffusion of the device was unproblematic. Problems with components were not unusual and these were subject to incremental modifications: the electrodes used in DBS could fray and turn on their axis (Siegfried and Shulman 1987); complications arose from lead implantation and the power source could fail (Shatin et al. 1986). In the 1980s some of these difficulties were overcome. Lithium batteries enabled the production of a neurostimulator that could be implanted for several years while providing the necessary level of continuous stimulation. New neurostimulator leads were produced based on endocardial (pacemaker) leads, and a device was created that could be programmed via a wireless console programmer. Again, these incremental developments were the result of technology transfer from the ongoing financially lucrative improvements associated with the pacemaker (Shatin et al. 1986).

Additionally, it is likely that the diffusion of the neurostimulator was linked to the success of the cardiac pacemaker in a broader sense. As Blume argues, the cardiac pacemaker gave 'the notion of an implantable device legitimacy and appeal' (Blume 2010, 34). It enabled clinicians to significantly improve the lives of a large body of patients and was heralded as a major advancement in modern medicine. This encouraged wider acceptance of implantable devices among the public, and no doubt encouraged those working in neurostimulation to emulate the success of this pacemaker.

The Advent of Medical Device Regulation

During this period, clinicians reported that many of their patients were responding well to neurostimulation therapies. In 1978, Cooper stated there had been a clinical improvement in the majority of the 700 patients who had undergone cerebral stimulation for the treatment of either cerebral palsy or epilepsy (Cooper and Upton 1978). Orlando Andy reported that the results of stimulating the thalamus at high frequencies in nine patients with motor disorders were 'fair to excellent' (Blomstedt and Hariz 2010, 431); and Heath, reporting on a study of 38 patients, argued that those with depression, those with 'behavioural pathology consequent to epilepsy, and those with psychotic behaviour consequent to structural brain damage' responded well to neurostimulation of the cerebellum (Heath et al. 1980, 243).

Yet in this area prior to medical device regulation it was not exactly clear what constituted a 'clinical improvement' or a 'fair to excellent' outcome. In effect, the opinion of a clinician was sufficient to determine whether a medical therapy was effective, and thus whether or not the use of a medical device could be justified. This reflected the sentiment of the early twentieth century when 'efficacy' was considered 'a matter of opinion, not a fact' (Bodewitz et al. 1989). With the advent of medical device regulation, however, 'efficacy' was delineated as something that should be objectively verifiable, thus necessitating clinical assessment tools and outcome measures that would enable the quantification of a patient's response to an intervention. Here we will see that there were few such tools for the conditions being treated with neurostimulation, and subsequently, neurostimulation therapies such as DBS for chronic pain were not approved in the new regulatory climate.

In 1976 after seven years of political debate, the US Congress passed the Medical Device Amendments to the Federal Food, Drug, and Cosmetic Act, thus granting the FDA authority over all medical devices. This was in response to a spate of device failures: between 1960 and 1970, medical devices were implicated in 10,000 injuries and over 700 deaths, and between 1972 and 1975, over 22,000 potentially defective pacemakers were recalled by manufacturers (Foote 1978). The intention of the amendments was to provide a 'reasonable assurance of safety and effectiveness for all devices' (Foote 1978, 111). With pharmaceutical regulation, the gold standard for determining efficacy became the double-blind trial. With medical devices, which are often not amenable to double-blind trials, the FDA stated that efficacy would be determined 'on the basis of well-controlled investigations, including clinical investigations where appropriate, by experts qualified by training and experience' (Foote 1978, 115). Consequently, efficacy was rendered as something objectively verifiable: it would have to be demonstrable to an FDA-appointed panel of experts that were not directly involved in the treatment in question.

In regard to neurostimulation technology used in DBS and cerebral stimulation therapies, the FDA decided that clinical trials would be necessary before it could be marketed (Coffey and Lozano 2006). Medtronic, Avery, and Neuromed, the three manufacturers of neuro-stimulation technology, were offered time to perform the necessary trials and produce the required documentation. All, however, eventually declined. The probable reason for this is provided by Adrian Upton, a UK-based neurologist. A major problem in the application of neurosti-mulation, he stated, was the lack of standardised, quantifiable measures for determining the effectiveness of the treatment (Upton 1986). Pain, for instance, was especially difficult to measure. In a recent summary of neurostimulation treatments for pain, Coffey and Lozano refer to the:

> paradox of pain – its simultaneous reality and subjectivity makes the assessment of pain relief therapies susceptible to observer- or patient-related influences . . . Unintentional cues, learned responses, or knowledge that a treating physician . . . is conducting the assessment can affect how patients rate analgesic treatments (Coffey and Lozano 2006).

Irving Cooper's neurostimulation therapies, including his DBS thera-pies, were also problematic in this regard. Despite Cooper's belief that they 'yielded promising clinical results', the lack of uniform objective evaluations to quantify and measure clinical improvement meant the 'true benefit' could not 'be elucidated' (Rosenow et al. 2002).

DBS and other neurostimulation treatments were not amenable to clinical trials because the conditions being treated were not quantifiable: there were no generally accepted clinical assessment tools that could be used to demonstrate clinical improvement. The clinical trials that were now needed to demonstrate efficacy and safety were also expensive, and Upton stated that the costs of further development would need to be offset by a large market (Upton and Lazorthes 1987). At the time Upton specifically referred to Parkinson's as a potential market for neurostimu-lators. The problem with Parkinson's, he claimed, was an over-reliance on medications which meant that the promising findings of earlier neurostimulation treatments for movement disorders were being over-looked (Upton 1986).

Clinical Trials, Evidence, and Clinical Assessment Tools

By the mid-1980s neurostimulation was in a precarious position. On the one hand, there was the medical device industry with the manufacturing skill set necessary to produce the neurostimulator technology, and there were neurosurgeons with the expertise required to incorporate the technology into working therapies. On the other hand, the new regulatory climate had effectively put a halt on the dissemination of many neurostimulation therapies and the market for the technology was restricted. In effect, medical device regulation established the FDA as what Latour (1987) refers to as a centre of calculation. It was endowed with the responsibility of assessing efficacy and safety of therapeutic interventions, and as a result it became a gatekeeper, either preventing or permitting dissemination of new device-based therapies. A distance was created between the point of therapeutic intervention (the clinic or the research site), and the point at which efficacy and safety of those interventions are assessed (an FDA-appointed panel of experts). Such distance was deemed necessary to reduce the influence of bias on the assessment of an intervention and thus provide a level of protection to patients and consumers. Yet, as Latour points out, for a centre of calculation to function, this distance must be traversed; the two points must, somehow, be brought together. This can be achieved via the production of immutable mobiles: renderings of entities of interest that are capable of circulating between the two locations without losing their meaning in the process (Latour 1987).

Clinical assessment tools are a means of generating immutable mobiles and creating an equivalency of meaning between the sites of intervention and the FDA, thus enabling the latter to act as a centre of calculation. In the mid-1980s, such a tool was developed for one particular condition that was known to respond to neurostimulation: Parkinson's. This tool, along with several other developments, encouraged Medtronic to pursue regulatory approval for an application of their neurostimulation technology: DBS therapy for Parkinson's.

In the mid-1980s at University Hospital in Grenoble, France, a team led by Alim-Louis Benabid was using ablative therapies to treat cases of Parkinson's, dystonia, and a few psychiatric conditions that had failed to respond to drug-based therapies. The team was one of the specialist centres that had retained the stereotactic tools and skill set developed during the subspecialty's period of growth. For each patient, a stereotactic apparatus was used in conjunction with imaging technologies to identify the areas for ablation and to plan the surgical procedure. Additionally, intra-operative electrostimulation was used to ensure the correct area had been located, and like others before him, Benabid noted that higher-frequency stimulation could reduce some of the motor symptoms of Parkinson's. Benabid set about trialling chronic neurostimulation as a therapy in its own right. Importantly, Benabid and his team had been using neurostimulation to treat chronic pain and were familiar with the equipment and methods that would be required to provide chronic stimulation: 'We had the method. We had the electrodes. We had the stimulating leads' (Talan 2009, 41). From 1987 onwards Benabid used Medtronic equipment to trial DBS of areas within the thalamus to treat tremor in patients with either Parkinson's or essential tremor, some of which showed complete relief (Benabid et al. 1987). While Benabid and his team were repeating what others had done a decade earlier, a particular conjunction of circumstances meant that Benabid's work was to have a far greater influence.

First, Benabid's team coupled DBS to Parkinson's at a time when clinicians were looking for a surgical alternative to levodopa-based therapies. Clinicians were becoming aware that while medications such as levodopa are initially effective in managing the symptoms of Parkinson's, they lose their effectiveness in the long run. By the mid-1980s, a reservoir of severely affected Parkinson's patients with symptoms no longer adequately managed with medications had emerged. Alternative therapies were needed, and neurosurgeons were beginning to revisit pre-levodop-era stereotactic surgical procedures (DeLong 1990; Laitinen et al. 1992).

Secondly, the accidental discovery of a neurotoxin led to the production of primate models of Parkinson's. The resulting studies enabled Benabid to consolidate particular areas deep within the brain as effective targets for DBS. In two separate incidents (1976 and 1983) recreational drug users inadvertently manufactured and ingested a compound that left them with severe Parkinson's disease-like symptoms. The substance was identified as MPTP (1-methyl-4-phenyl-1,2,3,6-tetrahydropyridine), and an autopsy later revealed that it has destroyed the dopamine-producing cells of the substantia negra, the same area that degenerates in Parkinson's (Porras et al. 2012). Subsequently, MPTP was used to create the first non-human primate models of Parkinson's, enabling new avenues of research into the underlying pathology; research that would have been unethical on afflicted human subjects. One such avenue of research produced a model of a pathological chain of neural activity in which the STN and the GPi are overactive (DeLong 1990). By surgically ablating these areas, researchers at Johns Hopkins University noted that they could reduce the induced Parkinson's symptoms in primates (DeLong 1990). The STN had first been identified as an effective target in late 1960s (Nashold and Slaughter 1969), but the resulting model now provided a scientific rationale, prompting Benabid to direct his attention to the area as a target for DBS.

And thirdly, Benabid and his team coupled DBS treatments to Parkinson's at a time when the disease could be quantified. In 1987, a consortium of movement specialists established the Movement Disorder Society and produced the Unified Parkinson's Disease Rating Scale (UPDRS). Their intention was to create a comprehensive and flexible system that would replace the numerous and idiosyncratic scales being used at various Parkinson's research sites (Fahn and Elton 1987). The variability of the scales in use at the time made comparative assessments difficult: the unified system would standardise clinical assessment across centres (Goetz et al. 2003). The UPDRS has five parts, each using a scale system to determine the severity of particular Parkinson's symptoms, including mentation, behavior and mood, speech and swallowing, facial expression, tremor at rest, rigidity, and finger tapping. For each symptom a number from 0 to 4 is used to assess severity (0 being normal or unaffected, and 4 being the most severe), and an overall score for each of the five parts of the UPDRS can be assigned to the patient. The severity of a patient's Parkinson's, therefore, can be represented with a series of numbers. In order to ensure that these numbers are equivalent across contexts, the Movement Disorder Society produced a teaching videotape specifically designed to aid new researchers and those conducting multi-centre trials (Goetz et al. 1995). The resulting equivalence would enable patients to be compared before and during treatment, across research centres, and would thus permit the calculations required to determine efficacy.

Given there was now a considerable demand for surgical treatments and that a tool had become available to quantify Parkinson's, it is not surprising that Medtronic enthusiastically aligned themselves with the Grenoble-based French team. In the early 1990s Benabid presented his results to Medtronic. Engineers at Medtronic had recently conducted studies to assess the use of their stimulation technology to manage pain, but these were abandoned due to the lack of any definitive results (The findings were eventually published in Coffey 2001). Benabid's work illustrated that the same technology could be used to treat Parkinson's and that the results could be demonstrated to regulators: 'Changes in movement are pretty obvious... pain is something that is not so obvious' (Medtronic engineer, quoted in Talan 2009). Subsequently, the highly regarded Benabid was employed by Medtronic to design international, multicentre clinical trials assessing DBS for the treatment of Parkinson's. The STN or the GPi were to be tested as target areas, both of which were supported by the newly developed model of deep brain function. The trials were funded by Medtronic and the UPDRS was used in all sites. Over the next few years, 113 people with Parkinson's and 83 with essential tremor were involved in the trials (Talan 2009).

In a laboratory trial, complexity is elided by having a few metabolic parameters 'stand in' for health; parameters that can be measured, counted, and used in the construction of factual statements. Similarly, this purification is required in clinical trials involving bodies. The UPDRS (like outcome measures in general) enabled the purification and inscription required to 'construct' factual statements regarding the efficacy of DBS. Because the UPDRS was adopted in all of Medtronic's DBS clinical trials, all patients were subjected to the same standardised regimes of examination, quantification, and comparison. Each patient, regardless of their unique personal histories or social context, was rendered as a set of comparable numbers representing the severity of their symptoms.

This quantification and the elision of messy and cumbersome personal detail, therefore, had three important functions. First, it enables calculation: it permits each participant to become a nexus linking the DBS technology to a clearly delineated region of the brain in a manner that could be clearly measured: the effect of DBS on particular regions of the brain the STN or the GPi could be determined by noting and comparing the numerical changes associated with each body. Second, these numerical renderings of the impaired body are mobile: they can be collected, pooled together, charted, graphed, compared, and computed, and these resulting inscriptions can then be circulated as 'proof' or 'evidence', with much more fluidity than fleshy bodies. These mobile numerical inscriptions are also immutable: they hold the same meaning across particular centres, enabling the establishment of a common language. Thirdly, as Porter has made quite clear, by eliding personal detail such renderings are imbued with an authority resulting from their supposed objectivity (Porter 1994). The UPDRS, therefore, was an essential part of an apparatus for producing facts in an era when 'efficacy' is institutionally deemed as something objectively verifiable. In effect, through a process of purification and inscription, the UPDRS created the immutable mobiles that bridged the distance between the point of treatment (the clinic) and the point of assessment (the FDA) that had been created with the advent of medical device regulation.

In March 1997, UPDRS-derived data were presented to the FDA's Neurological Devices Panel Advisory Committee by consultants managing the trials on behalf of Medtronic. The intention was to gain approval for the 'Medtronic 3382 DBS lead and the Medtronic ITREL stimulation system for the suppression of tremor due to essential tremor or Parkinson's; unilateral or bilateral' (FDA 1997). Slides of the results were shown to the panel and explained by a consultant who drew attention to both individual improvements in UPDRS and statistical analyses of overall UPDRS data. He stated that there was a statistically significant reduction in tremor and global disability, and that the efficacy of the treatment appeared to be greater than that of available medications. There was more to the panel hearing than the presentation of numerical data, however. A portion of the hearing was reserved for members of the public to voice their opinion. Four individuals spoke to the panel, all in support of the therapy. The testimony of these four speakers highlighted the day-to-day difficulties of living with a movement disorder, and the hope and expectation that had been invested in the DBS therapy. Two were representing patient advocacy groups and two were being successfully treated with DBS therapy as part of the clinical trial (three of the four had also been brought to the hearing by Medtronic).

At the end of the hearing, members of the panel expressed their initial impressions. All members believed the efficacy of DBS had been demonstrated. Some, however, said they were not convinced of its safety: although no major safety issues had been identified, potential adverse effects could not be ruled out. As one member put it: 'unless the side effect slaps you in the face or is quite profound, you may not find subtle effects like the neurological changes' (FDA 1997). As a result of the panel's findings, the FDA formally approved the use of unilateral DBS for the treatment of essential tremor and Parkinson's. For the latter, however, it could only be used in patients with severe tremor, due to uncertainty regarding possible adverse reactions. Medtronic continued to sponsor trials exploring the longer-term effects of stimulation, and in 2002, confident that sufficient evidence of safety had been demonstrated, the FDA approved DBS for more general cases of Parkinson's. In 1998 the same clinical trials were used by Medtronic to gain the CE mark for their technology, thus permitting the use of DBS to treat Parkinson's within the European Union.

A History of DBS: Key Themes

Within ten years of being approved for the market within the USA over 40,000 individuals had been treated with DBS for Parkinson's or essential tremor (Talan 2009), and DBS had gone from being a marginal therapy to an 'effective', 'standard and accepted treatment for Parkinson's disease' (Montgomery and Gale 2008). Currently, people with Parkinson's can undergo DBS therapy at specialist centres in North America, Australia, and New Zealand, and much of Europe including the UK. In their 2014 annual report Medtronic highlighted their 1.9 billion USD in revenue from their Neurmodulation division, a large proportion of which is driven by sales of the DBS technology in both Europe and the USA (Medtronic 2014). DBS for Parkinson's, then, has proven highly lucrative for Medtronic. By forming close alliances with key clinician, Medtronic has facilitated the adoption of their technology into other treatments. While Parkinson's is by far the most common DBS indication, a small number of patients have now undergone DBS for the treatment epilepsy and OCD and depression, and over a thousand have received DBS for primary dystonia. Again, the flexibility of the technology has enabled it to be easily adapted to other treatments: as one financial analyst recently put it, 'deep brain stimulation provides a stimulating market . . . because with a single platform, companies can address several diseases with large populations' (Stuart 2012). Some estimates proclaim that by 2020 the global neuromodulation market will be worth over 6 billion USD (MarketsandMarkets 2015).

Commercial interests, then, have clearly been important in shaping the innovation trajectory of DBS. So have professional interests: the flexible DBS technology has had strategic utility for clinicians, particularly neurosurgeons. With the 'rebirth' of DBS neurosurgeons with training in stereotactic techniques are once again gaining access to a pool of patients with conditions that are inadequately managed with medications. As Gildenberg puts it, in the early 1990s stereotactic surgery was: 'the realm of a relatively small group of subspecialists'. By the end of the 1990s: 'more stereotactic surgery was being practiced by more neurosurgeons than ever before, and stereotactic techniques made inroads to become needed skills for every practicing neurosurgeon' (Gildenberg 2000, 309). Importantly, dissemination has also been driven by an increasing realisation of the limitations of drug-based therapies. As Ackerman points out, neurologists are far more inclined to recommend surgical therapies to patients in the present day. Indeed, in the majority of present-day centres offering DBS, patients are managed by teams that include both a neurosurgeon and a neurologist (Ackerman 2006, 111).

As this historical overview has shown, the development and stabilisation of DBS therapies was also contingent upon the co-development and diffusion of standardised methods of rendering the affected body intelligible. By eliding complexity and foregrounding specific phenomena as 'significant', clinical assessment tools have become an essential technology in the innovation process. The UPDRS, for example, has been adopted throughout Europe, the USA, and elsewhere, and is now considered the gold standard reference scale for Parkinson's. Dystonia, like Parkinson's, lends itself to quantification techniques, and in the mid-1980s a dystonia-specific clinical assessment tool was produced. The Burke–Fahn–Marsden Dystonia Rating Scale (BFM), like the UPDRS, uses a series of numbers to indicate the severity of symptoms (Burke et al. 1985). The BFM has become the standard clinical tool for measuring the severity of dystonia in neurology, and it was used in the clinical studies that led to regulatory approval. These are studies involving adults with primary dystonia, and in 2003, Medtronic DBS technology was fully approved for the treatment of dystonia within the European Union and partially approved within the USA. Similarly, Medtronic sponsored trials assessing DBS for epilepsy and OCD have also adopted clinical assessment tools that produce computable, mobile, impersonal renderings of the patient. Such tools have become vital in the era of medical regulation; indeed, the lack of such a tool for a given illness may discourage innovators, as has been the case with the DBS to treat chronic pain. As we shall see, the necessity of such tools has implications for the PMDS.

Implications for the PMDS

It is this conjunction of circumstances – brought about by an interweaving of commercial and professional interests and parallel technological developments – that brings us to the PMDS. The context of the PMDS is characterised by a significant demand for new therapies to treat neurological illnesses that cannot be managed with medications. Neuromodulation is increasingly being championed as an option. By the mid-2000s, several adult-focused centres in the USA and Europe had provided DBS to a small number of children with otherwise untreatable primary dystonia. The effectiveness of these initial paediatric cases encouraged Dr Martin to establish the PMDS team in 2005, three years after Medtronic has received regulatory approval to market DBS as a treatment of dystonia. Medtronic continues to be a dominant player in the neuromodulation market: the team uses Medtronic DBS technology and regularly communicates with Medtronic representatives, and Dr Martin has worked as a consultant for the company. As with other centres providing DBS therapies, the PMDS involves a close alliance between neurology and neurosurgery. In this regard, it is representative of the new collaborations that have formed around neuromodulation interventions.

Additionally, the PMDS inhabits a context in which they are expected to be able to demonstrate the effectiveness of their paediatric DBS therapy by producing objectively verifiable, quantitative data. The necessity of this is, of course, a reflection of the rationalisation of healthcare, of which the EBM movement is a significant component (Porter 1995; Wehrens and Bal 2012). Rationalisation has involved a move towards pre-defined processes aimed at improving efficiency, an emphasis on quantitative over qualitative characteristics, and creating uniformity over multiple sites (Ritzer 2014). Gatekeeping governance mechanisms such as regulatory frameworks and HTA agencies (such as the UK's National Institute for Health and Care Excellence) have been a major force in the entrenchment of rationalisation in healthcare. The use of DBS for dystonia has received regulatory approval, but this was based on studies involving adults with primary dystonia: it has yet to be widely appreciated as an acceptable therapeutic option for children with secondary dystonia, which is more complex and more difficult to manage. As a pioneering service, the PMDS is required to produce evidence that would enable them to appraise the effectiveness of their service, which would serve to convince the wider neurological community that the therapy is worth pursuing. As we have seen earlier, there are standard assessment tools that can help them to produce this evidence, namely the BFM that was used in the original studies of DBS for adults with dystonia. However, as we will see in Chapter 7, there are limitations with this tool: it does not, PMDS team members argue, capture improvements that patients themselves feel are meaningful.

The entrenchment of rationalisation via governance mechanisms has created an additional reimbursement challenge for the PMDS. Initially, the PMDS received financial support via an institutional charity grant.

The long-term sustainability of the service depends on their ability to cover costs through existing commissioning arrangements within the NHS. As we will see in the next chapter, many of the team's activities (such as consultations with patients) can be supported through the NHS's existing payment system for specialised paediatric services. Other activities, however, such as the costs of the DBS technology and its surgical implantation, are not provided for within this system. Instead, these have had to be commissioned by the regional commissioning groups within the NHS, each of which has its own specialised services (i.e. non-routine) commissioning policy, and each of which only funds services for patients living in their jurisdiction.[4] The decisions of these groups are informed by available evidence on the cost-effectiveness and clinical benefits of the treatment in question. Given the relative novelty DBS, such evidence (for both adult and paediatric) is sparse. For some groups, the available evidence is sufficient to justify commissioning the therapy, yet for others, the same evidence pool has been deemed insufficient. One group has declared, for example, that: 'the use of DBS for non-Parkinson's tremor, dystonia and pain disorders is not supported due to a lack of evidence' (East Midlands Specialised Commissioning Group 2011, 2).

This divergence in commissioning policy has received sustained critical attention. The Dystonia Society described it as a 'postcode lottery' (The Dystonia Society 2010), and it became the subject of a Channel 4 News item (4 News 2012). The Dystonia Society enrolled Lord Macdonald and several MPs including Stephen Dorrell, the chair of the Health Select Committee, in a campaign pushing for a National Commissioning Policy that would force all regional commissioning groups to align their position on DBS. The campaign appears to have been a success: now, all adults who meet eligibility criteria will receive funding for DBS for dystonia, Parkinson's, or essential tremor.

---

[4] The total cost of DBS for paediatric dystonia is around £80,000 per patient over ten years. Approximately, £35 000 of this covers hardware and surgical costs, and the remainder covers pre- and post-surgical assessments. The cost of adults is around £5000 less over ten years. This is because adults undergo less rigorous, and thus cheaper, regime of assessments (Yianni et al. 2005; Medical Services Advisory Committee 2008).

As yet, however, there is no national commissioning policy for paediatric DBS, and the PMDS can only accept referrals if they are accompanied by a letter from the patient's Commissioning Group guaranteeing payment for the DBS implantation. In the past, commissioning groups (then called 'Primary Care Trusts') within the East Midlands area had refused to provide funding, and several families wishing to peruse DBS with PMDS had moved postcodes in order to fall within another jurisdiction.

Reimbursement, then, has represented a challenge for the dissemination of DBS for dystonia within the UK. Through the manoeuvring of the Dystonia Society (a patient advocacy group) and their enrolment of political allies, this challenge has to some degree been dealt with. For the PMDS, however, the current uncertainty around reimbursement for paediatric DBS is another reason why it is necessary to collect comprehensive, objectively verifiable, and as we will see, patient-centred evidence.

Summary

In this chapter we have delved in the history of DBS and explored some of the factors that have shaped its development and dissemination. To summarise, I've suggested that we can see DBS as having its origins in ablative therapies such as the Montreal Procedure for treating epilepsy. Given the huge demand for surgical treatments for psychiatric and movement disorders, and due to the success of the stereotactic apparatus, ablative therapies were conducted in many centres, bringing about a period of 'unrivalled' experimentation. From this has emerged much of the skills and knowledge and some of the technology that would later inform DBS. The introduction of pharmaceutical medications for prevalent movement and psychiatric disorders drastically reduced the pool of patients for ablative therapy, and stereotactic surgery became a speciality practice in a few clinics worldwide. Meanwhile, the burgeoning cardiac pacemaker industry generated new technology platforms and vast sums of money that, with some creativity from neurosurgeons, spawned the neurostimulator. The dissemination of these neurostimulator in some treatments was halted with the advent of medical device regulation. In the mid-1980s, when there was a resurgence in demand for surgical treatments for Parkinson's, and when a tool for quantifying Parkinson's became available, Alim-Louis Benabid and his team initiated what we might call the current era of DBS. Medtronic-sponsored clinical trials have led to regulatory approvals for several indications, dependent on the existence of appropriate clinical assessment tools for generating evidence. More recently in the UK, commissioning policy has presented some hurdles for those wishing to access DBS for dystonia.

These are the historical circumstances which to some extent bear on the activities PMDS: some of the knowledge which informs the team's activities, and some of the tools and technologies utilised by the PMDS have their origins in these historical developments. We can of course describe these as being constituent elements of the PMDS *proto-plat-form*. However, the team also represents a stark departure from these earlier developments. The team represents a point at which the innovation trajectory of the DBS technique interweaves with another development in healthcare: the patient-centred medicine movement. It is this particular conjunction that is the focus of the next four chapters, because it represents a potentially significant transformation in the way in which movement disorders – and indeed the patient – are being rendered intelligible in clinical work. We now zoom in and explore how the team is managing specific challenges relating to the adoption and implementation of DBS in paediatric neurology: coordinating multidisciplinary teamwork (the next chapter), identifying suitable candidates for DBS, managing patient expectations, and measuring clinical outcomes. In each chapter I draw on my observations of PMDS teamwork and interviews with team members, and we examine proto-platform elements that have emerged as team members have sought to manage these challenges. These are, as we will see, elements that reflect the values of patient, and they collectively entail the enactment of a broad clinical gaze.

4

Multidisciplinary Teamwork

The PMDS team very kindly allowed me to observe them at work for a period of just over a year. During this time I was able to sit in and observe their team meetings, I was able to watch them working with patients and families, and was also able to speak with each team member individually and discuss with them things that I had seen. The advantage of this is that it enabled me to see first-hand the types of challenges they encountered and how they managed these challenges, without having to rely solely on their own accounts of what they were doing.

My observations of team meetings were especially useful in this regard, and it was during my very first team meeting observations that the first challenge we explore became apparent. Several features of the meetings caught my attention; features that initially struck me as peculiar, but which I soon realised were characteristic of much of their work. The first of these was the way in which members of the team referred to, and spoke about, their patients. While there was some talk about MRI images, brains, the basal ganglia, electrodes, and stimulation parameters, much more time was devoted to talking about a child's progress at school, their relationship with their siblings and parents, their mood, and their ability to access computers, iPads, or other assistive technologies. Importantly,

© The Author(s) 2017
J. Gardner, *Rethinking the Clinical Gaze*, Health, Technology and Society, DOI 10.1007/978-3-319-53270-7_4

such considerations had a bearing on the decisions that were made during team meetings. The patients who were being discussed, then, were not the impersonal biomedical entities that have been the subject of considerable critical attention from social scientists. Given the various professional backgrounds of those involved in the team, I should not have found this surprising: proponents of 'multidisciplinarity' (or 'interprofessionalism') in healthcare have argued that it leads to a more 'comprehensive', 'patient-centred' approach.

This brings me to the second feature of the meetings that initially struck me as peculiar: the seemingly significant proportion of meeting time that was devoted to discussing what could be called administrative issues. Much of the talk pertained to aligning the weekly schedules of various team members, encouraging team members to complete and circulate reports, planning how to secure necessary resources and clinical working space from the hospital, and bemoaning the length of the meeting itself (which was seldom completed within the allocated one-hour slot). My understanding of these discussions is that they reflect the difficulty of coordinating teamwork; a task which is no doubt complicated by the multidisciplinary nature of the team, but is essential, proponents argue, if a 'comprehensive', 'patient-centred' approach to healthcare is to be achieved.

So, from very early on during my observations I became aware that multidisciplinary teamwork is not easy, and that certain strategies are used within the PMDS to coordinate the activities of various team members. The product of these coordinating activities is a particular rendering of the patient: a patient who is understood as more than just a biomedical entity – at least during team meeting discussions. In this chapter, I flesh out this relationship between DBS, the techniques for managing and coordinating multidisciplinary teamwork, collaboration, and a specific way of 'enacting' the patient. We will see that there is a particular rationale for having a multidisciplinary team to implement DBS; a rationale which reflects prominent neurodevelopmental understandings of childhood. We also see that this neurodevelopmental 'way of understanding' is reflected in various elements of the context within which the team works, such as the hospital architecture and the payment structures for paediatric services. We interrogate how these elements (along with several other features of PMDS context) facilitate multidisciplinary teamwork, and importantly, we also see how they collectively enable what I will define as a broad clinical gaze. Towards the end of this chapter, I suggest that such elements, and indeed the structure of the team itself, can be seen as constituting part of a socio-technical proto-platform for DBS. Before we do this, however, I provide some important context to the discussion by providing some background on the challenge of multidisciplinary service provision in healthcare.

The Challenge of Multidisciplinarity in Healthcare Settings

For the past two decades health professionals and policy makers have actively promoted multidisciplinary team-based service provision within healthcare. Advocates have argued that multidisciplinary (or 'interprofessional') teams are an important means by which safe, efficient, and patient-centred outcomes can be accomplished (Department of Health 2000; Kennedy 2001). Teams that include individuals from different professional backgrounds, it is argued, can tap into a range of disciplinary perspectives and are more likely to provide a comprehensive assessment of a patient's needs, and are thus better suited to help the patient decide upon the most appropriate course of action. As Finn and colleagues argue (Finn et al. 2010), a pro-teamwork discourse is a defining feature of healthcare reform, particularly in the UK. This is illustrated in various UK Department of Health documentation, such as *NHS: Next Stage Review: A High Quality Workforce*, which states that achieving the NHS's goal of delivering 'high quality care for patients and the public... requires an effective team of professionals across clinical, managerial and supporting roles' (Department of Health 2008, 8).

Over the past few decades this pro-multidisciplinarity perspective (which also aligns with the 'patient-centred healthcare' discourse) has become encoded in UK legislation, and this, in turn has had a real impact on levels of interprofessional collaboration. This has particularly been the case in the provision of primary, community-based care for those with psychiatric illnesses, chronic illness, people with disabilities, and the elderly (D'Amour et al. 2005). The Health Care Act 1999, for example, created 'pooled budgets' for health and social services to encourage local authorities and health authorities to merge services. This enabled the formation of 'rapid response teams' that could provide emergency care for people at home, and 'integrated home care teams' to help people live more independently, both of which might consist of nurses, care workers, social workers, therapists, and GPs. Such policy initiatives have helped to foster an environment in which it is common for occupational therapists, for example, to liaise with social workers, physiotherapists, and doctors. Child Development Teams (CDTs) are a good example of this form of collaboration. There are currently over 300 CDTs in the UK, one for each district. The intention of these teams is to assess, diagnose, and help manage children with developmental delays and disabilities in a community setting. Typically, these teams involve occupational therapists, physiotherapists, a paediatrician, a speech therapist, and a social worker, and most teams also have access to a child psychiatrist (McConachie et al. 1999). As we shall see, the PMDS team replicates this mode of operation – indeed, several members of the PMDS team had worked within such teams before their time with the PMDS.

Generally teamwork within hospital settings has tended to involve a less diverse array of professionals. (Hospital medicine, as May and colleagues (May et al. 2006) point out, has tended to be less interested in the 'comprehensive' approach to dealing with illness.) Surgical teams of course are commonplace, and are orientated towards supporting the actions of the lead surgeon. Critical care is also provided by multidisciplinary teams, generally comprising nurses and doctors with various clinical specialisms. And since the 1990s, multidisciplinary teams, also comprising doctors and nurses with various specialisms, have emerged as the predominant method of delivering oncology services (NHS National Cancer Action Team 2010).

Various authors have highlighted the difficulties of establishing and working within interdisciplinary teams in healthcare settings.[1] Hardy and Turrel (Hardy and Turrell 1992), for example, identified five categories of

---

[1] Strictly speaking, interdisciplinarity refers to attempts to integrate disciplinary perspectives within a team, while multidisciplinarity refers to the coordinating of disciplines which themselves remain distinct. But, in practice, as Barry and colleagues (Barry et al. 2008) have argued, the terms tend to be used interchangeably.

barriers to joint working. These included: *structural* and *procedural* issues, such as differences in planning procedures and budgetary cycles that may prevent, for example, effective collaboration between health and social services; *financial factors* such as the cost of establishing and maintaining the appropriate administration and communication infrastructure; *status and legitimacy* issues that can exist between the different authorities responsible for providing services (such as NHS and appointed local authorities); and *professional issues*, such as the difficulty in aligning diverse professional perspectives, expertise and skills, and the problems associated with professional self-interest. The last of these – professional issues – has received quite a bit of critical attention. As Pietroni has noted (Pietroni 1992), the language and values that characterise disciplines can hinder the ability of interprofessional communication. There can be, in other words, a degree of incommensurability between different professions, each of which represents a distinct community characterised by particular expertise, language, and tools (Fournier 2000). Other authors have noted that traditional hierarchies of professional authority can also impede the intended benefits of teamwork from being achieved. This has been noted within multidisciplinary critical care teams, for instance, in which doctors tend to have a far greater influence on clinical decision-making than nurses (Coombs and Ersser 2004; Coombs 2003). During their ethnographic studies of several clinical sites, Coombs and Ersser (2004) noted that a patient's course of action was generally decided upon by doctors based on their biomedical understanding of the body. Nurses, who had often acquired a very detailed knowledge about the patient's family and the patient 'as a person', had comparatively little input. This, Coombs and Ersser argue, is indicative of a 'medical-hegemony' in clinical decision-making (Coombs and Ersser 2004, 245).

There are, then, considerable challenges in multidisciplinary team-based health service provision. Such teams are an intersection of professional worlds, each characterised by distinct tools, understandings, and practices. According to current policy discourse, this diversity is necessary for effective service provision, but in actual clinical practice, this diversity threatens to undermine the ability of such teams to function effectively. The existence of multidisciplinary teams, then, is dependent upon strategies that both preserve diversity and coordinate diverse activities towards the achievement of the aims of the team as a whole. In what follows, I explore some of the strategies that bind the various members of the PMDS team together and coordinate their diverse activities towards the provision of a DBS service. Specifically I explore several key 'binding' elements: (1) the built environment of the hospital which was intentionally designed to foster multidisciplinary activities; (2) the creation and circulation of a schedule which coordinates the activities of team members; and (3) the weekly team meetings, which provide both a forum for prompting individuals to align with team strategy and a structured space for multidisciplinary decision-making. First, however, we explore why the PMDS team was established as a multidisciplinary team in the first place.

The Clinical Justification for a Multidisciplinary PMDS

The composition of the PMDS team reflects a particular understanding of motor system development. According to this understanding, a new-born child has a motor system with latent potential but which is largely unusable. It is through constant sensing of, and interaction with, the surrounding material environment that a child's motor system gradually becomes highly specialised and adapted, providing the child with a capacity to move the body and negotiate space ('fine' and 'gross motor function'). Importantly, this stimulates cognitive development: it involves, for example, the development of spatial awareness and the ability to anticipate the consequence of particular bodily actions. Generally, a child's progress is indicated by well-known developmental milestones: crawling, sitting upright, standing, learning to walk, and so on. If development is unimpeded, the child will learn how to use their body and obtain the necessary cognitive skills to conduct a vast range of culturally mediated and meaningful activities: riding a bicycle, using a keyboard, brushing teeth (often referred to as 'tasks of daily living'). According to this neurodevelopmental understanding of childhood development, dystonia, like other childhood physical disabilities, is perceived to be problematic because it inhibits the child's ability to engage in the social and physical interactions and educational opportunities that are considered necessary for healthy development.[2] In effect, it 'locks' the body, rendering it insensitive to the physical environment and unresponsive to the child's will. Elements of the child's motor system will fail to become honed and adapted, and this may consequently stunt some aspects of cognitive development.

By targeting very specific areas of the brain with electrical stimulation, DBS promises to 'unlock the body', thus enabling a child (supported by therapists) to engage in the interactions that are necessary for healthy development and meaningful living. As the Clinical Research Fellow states:

> The way I see it is that the DBS unlocks dystonia so the therapists can get in and give you the functional recovery . . . it allows you to get some more intensive therapy. (Clinical research fellow, interview).

Accordingly, the PMDS was composed by Dr Martin to include a set of professionals with the skill sets to identify both how dystonia may be hindering various aspects of development, and how such hindrances might best be managed: a physiotherapist with the ability to examine the impact of dystonia on fine and gross motor function; an occupational therapist to examine the impact on the child's ability to conduct 'activities of daily living' tasks; a psychologist to assess cognitive skills; and a speech and language therapist to examine talk and verbal communication, swallowing, and feeding (Clinical research fellow, interview). Clinical exigencies dictated which other professions would be included in the PMDS. Neurologists are obviously important due to their knowledge of the central nervous system and medications: the former is drawn upon to identify the DBS target and plan the implantation procedure, and the latter is necessary to advise on patients' medication regimes.

This neurodevelopmental understanding of childhood development, and the corresponding belief that a multidisciplinary approach is the appropriate

---

[2] This neurodevelopmental model aligns with what Meloni (2014) has referred to as the 'ultra-social brain' of contemporary neuroscience, in which the brain of the individual is conceived as a multi-connected social entity that is affected, and to some degree shaped, by the individual's social context.

method of service provision for children with disabilities, are prevalent within the UK. They have had a notable impact on child health and social welfare policy, and indeed this has enabled establishment of the PMDS as a multidisciplinary team in several ways. First, as stated earlier, multidisciplinary Childhood Development Teams are the common means of delivering health services to children with severe disabilities in community settings throughout the UK. Within these teams, it has become commonplace for occupational therapists, physiotherapists, speech and language therapists, and doctors to work together with clients (McConachie et al. 1999). The PMDS physiotherapists and the occupational therapists had worked in such settings prior to joining the PMDS team and were familiar with each other's professions. In this respect, the PMDS team structure mimics a model of service provision that has been in existence for several decades, albeit in community setting. Second, both the Dystonia Society UK (Dystonia Society 2014) and the existing NICE guidelines on DBS (NICE 2006), recommend a multidisciplinary assessment process for DBS (Gimeno and Lin 2016). The PMDS has drawn on both of these to justify its structure. Team members have also stated that the PMDS multidisciplinary structure is intended to reflect the World Health Organisation's International Classification of Functioning, Disability and Health; a classificatory system which aims to capture the physical, psychological, and social aspects of disability.

And third – and very importantly – current payment structures within the NHS reflect the belief that a multidisciplinary approach is the preferable mode of service provision for children. Payment structures within healthcare systems (i.e. how hospitals and other providers are reimbursed by insurers or the State for the services they provide to patients) are generally designed to encourage the efficient use of resources and to remove highly variable standards of care. As we saw in Chapter 2, such structuring payment systems can discourage technology adoption: providers (i.e. hospitals) will be reluctant to provide diagnostic or therapeutic services if they can't be accommodated within pre-set payment amounts (tariffs), even if they result in cost savings for the health system as a whole (Llewellyn et al. 2014). Multidisciplinary services that include physiotherapists and occupational therapists can be expensive; in many cases too expensive for providers, even in disease areas where such an approach might be considered optimal. Within the NHS, however, the tariffs for an intervention for a child are generally higher than the equivalent intervention for an adult. For example, as of 2013 the base tariff for adult's first attendance to a neurology outpatient consultation is £225. For a child, the same service has a tariff of £400 (Department of Health 2012). Additionally, a hospital that delivers a specialised service intervention for a child will receive 'top-up' of between 44% and 64% (NHS England 2014). This, according to one interviewee I spoke to (an ex Clinical Director at the Children's Hospital), is based on the recognition that a child may need the support of a greater range of professionals. This commissioning policy has, then, reduced some of the financial hindrances to multidisciplinary service in paediatric service provision. Indeed, along with the commitment by most Regional Specialised Commissioning Groups to fund DBS for children (as we explored in the last chapter) such payment structures have enabled the PMDS to become financially self-sustainable, and it has enabled them to employ their own administrator who, we will see, plays an important role in coordinating the activities of other team members.

The payment system has enabled the team to establish what might be called a 'comprehensive' patient pathway (see Fig. 4.1) within which team members will have considerable contact time with patients and families. Specific aspects of this pathway are explored in subsequent chapters, but here I will provide a brief overview: First, the families of new referrals are sent a questionnaire that explores the patient's level of disability, their medication regime, and their history of clinical interventions. The patient and supporting family members then attend a screening clinic at the hospital, during which one of the neurologists and another member of the team (generally the occupational therapist or a physiotherapist) undertake a preliminary assessment of whether or not the patient is a suitable candidate for DBS. If, initially, they are a good candidate, the patient will return at a later date to undergo a series of scans (MRI and PET imaging), an information session about DBS, and a regime of pre-surgical assessments, which may take several days to complete. These assessments are used to confirm that they are indeed a suitable DBS candidate and to produce a 'baseline' measurement of the severity of their condition. A range of features are assessed, reflecting the collective skills and expertise of

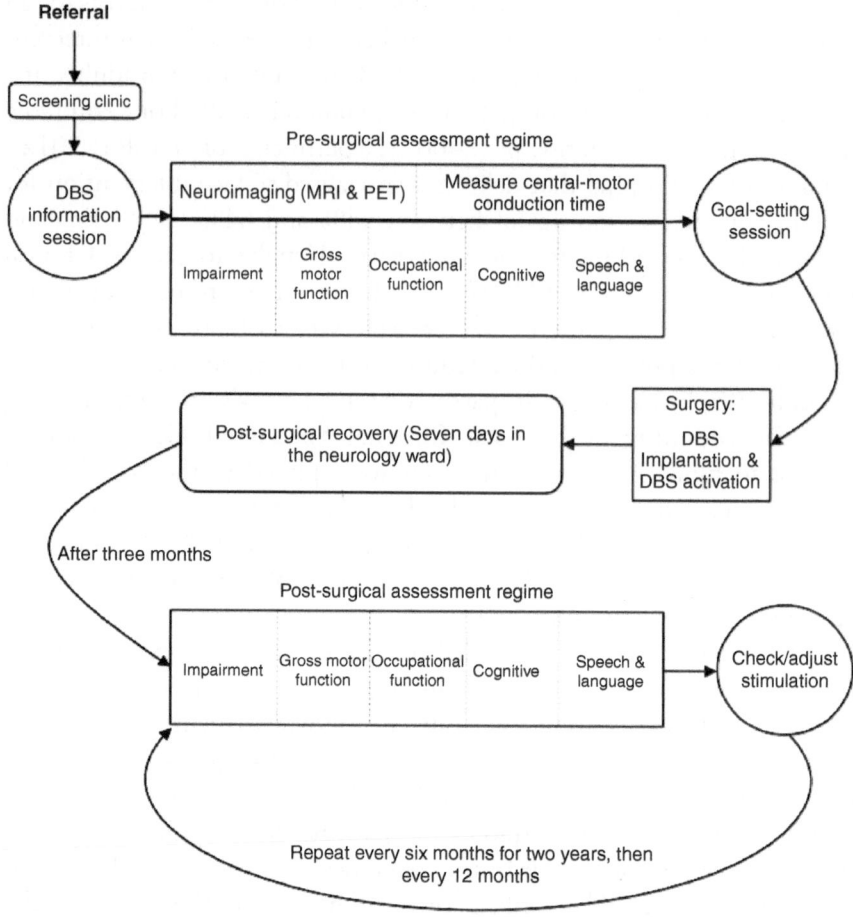

Fig. 4.1 The usual pathway for a DBS patient in the PMDS

the team as a whole (and indeed the neurodevelopmental model of childhood development):

- Impairment: the degree to which the central nervous system is affected by the motor disorder;
- Gross motor function: the degree to which the disorder affects the patient's ability to perform major movements, such as standing, walking, and sitting upright;

- Occupational performance: the ability of the patient to undertake common, day-to-day tasks, like brushing teeth;
- Cognitive ability: the ability of the patient to perform basic cognitive tasks;
- Speech and language: the ability of the patient to verbally communicate. This may also include assessing the patient's ability to chew and swallow food items.

Conducting these assessments with a patient and supporting family members may take a couple of days, during which some team members – particularly the therapists – may come to know a family quite well. Some aspects of this pathway, especially the DBS hardware, the surgical implantation, and the pre-surgical assessments, are paid for via the Specialised Commissioning payment, while other aspects are covered by existing tariffs for paediatric services. It should be noted at this point that this pathway is not, due in part to financial constraints, replicated by those few UK clinics providing DBS to adults with dystonia.

In addition to payments structures, the neurodevelopmental model and the championing of multidisciplinarity have also influenced the design of clinical spaces. In the following section we explore how the built environment in which the PMDS team is based has been designed to encourage multidisciplinarity, and importantly, we will glimpse how it actually influences their activities.

The Children's Hospital: Materialised Multidisciplinarity

Various authors have explored the relationship between medical knowledge, hospital design, and biomedical practice (see Prior 1988; Keating and Cambrosio 2003; Adams 2008). Prior argues that the spaces that constitute the built environment of a hospital should be seen as a social product: the physical form of a hospital, such as the layout of the beds and the location of amenities, embody culturally mediated understandings about the patient and disease, in addition to engineering considerations, aesthetics, and so on. And importantly, Prior argues, such spaces are not merely an inert backdrop to human interaction: their symbolic and material dimensions guide, restrict, and facilitate interactions, and thus encourage specific forms of biomedical practice. Drawing on the work of Foucault (1991), Prior argues that hospital architecture contributes to the production of particular forms of clinical work: it contributes to configuring clinical practices within, and thus to the enactment of health, illness, and the patient that such practices generate.

Using Prior's conceptualisation we can draw an interesting link between the architecture of the children's hospital and the multidisciplinary activities of the PMDS team. The design of the recently built hospital reflects the belief that multidisciplinary service provision results in better health outcomes for patients and, consequently, the built environment of the hospital itself *encourages* (but does not *cause*) particular forms of teamwork and multidisciplinarity. This can be illustrated with extracts from a discussion I had with several clinicians, one of whom was the Clinical Director of Children's Services at the time the hospital was designed and built. Here I briefly explore three aspects of the hospital: the open-plan office space, the ward design, and the greater hospital itself.

The Office Space

The office space in particular was designed to encourage a team-based approach. There is space for 150 clinical and administrative staff, and while it contains seven or eight enclosed office rooms, much of the space is open plan. The rationale for this is explained by the then Clinical Director:

> We were keen to try and bring teams together, because people had come from little boxy offices, all over the place, and often very isolated. What we did was to [have] areas for teams. So there's a neurology bay, a cardiology bay, a nephrology bay and so on.

When teams were moved into the office area that they had been allocated, they could then decide how exactly they wanted to use it:

> What some specialties did was put the secretarial staff, who are in their offices all day, in the closed offices, so they would have a bit more peace and quiet and get on with their audio typing and so on. Clinicians, who were coming and going all the time, were put in the open plan offices. Other specialties did it the other way round...My own view is that it worked best when the clinical teams were in the open plan area, because they saw each other more, they could talk to each other more, they could provide support.

The PMDS team arranged themselves so that all members, including the administrator, occupy roughly the same area of open-plan office space. Indeed, it was important to Dr Martin (team leader) that it remained this way in the face of a reshuffling that was occurring to accommodate additional hospital staff. In one team meeting, for example, Dr Martin told the team that 'I really don't want to be scattered around the place', and it was noted in the minutes that a hospital administrator would be contacted to ensure that this was avoided.

For many team members, the seating arrangement did indeed enable them to easily consult one another (and other clinicians) about cases, and according to one team member it also helped to facilitate a positive social atmosphere:

> It really does encourage people talking to each other and because you're more visible, it encourages ease of access. It breaks down the barriers of, 'Oh I've got to go and knock on someone's door.' I think it does pull the team together...Socially it's lovely because we all get to chat with each other. It's not quite as productive as it could be, for that reason. (Clinical Research Fellow, Interview)

It appeared that a particularly important benefit of this open-plan arrangement was that it enabled team members to constantly check and co-ordinate their schedules: it facilitated, in other words, the planning of collective action. The administrator (who, as I will illustrate further on, is responsible for producing a weekly team schedule) felt that it was particularly helpful for her work. Here she recounts a previous temporary seating arrangement where team members were more scattered:

> Some of us were in every single lobby scattered around. I think we work – I certainly work – a lot more efficiently [when] I can see all of them. If I had a quick question or someone's on the phone and I've got a quick query – I would have to [say], 'Oh let me write that down and when I can find that person then I'll call you back.' It's a lot easier if I just access my team, have them sitting around me. (Administrator, Interview)

The Hospital Wards

During the design of the hospital, focus groups consisting of clinicians from a range of professions (nursing, physiotherapy, occupational therapy) were consulted on the ward layout (Clinical Director, Interview). One result of this is that the wards were designed to be flexible spaces, in which a range of varied services could be brought to the bedside of the child, rather than the child having to be transported around different departments of the hospital. As the Clinical Director stated:

> So that's thinking along a multidisciplinary route, you're acknowledging that the child actually may need to be seen by a number of different clinicians...And the way the building is designed, the services come to the child. Any bed should be able to support any type of service...wherever possible, the services should be taken to the child rather than the other way round (Clinical Director, interview)

Creating such a flexible 'child-centred' space, the Clinical Director also noted, required investing in a particular technical infrastructure. 'A lot of money', was spent on portable equipment such as x-ray machines and devices such as specialist power pendants:

> A German manufacturer . . . could provide these pendants where the power sockets are on columns that come down out of the ceiling, so that you can put the bed in the middle of the space and people can actually look after the child round all four sides of the bed (Clinical Director, Interview).

The flexibility of the ward space, then, is enabled by a configuration of various technical and material elements. Clinical tools are sufficiently mobile to move from one bed to another, and the ward itself provides sufficient space for a team of clinicians to access the bed. Obviously, other medical and social considerations influenced the design of the hospital. Limiting cross-infection and promoting interactions between children, for example, were two concerns mentioned by the Clinical Director

After the surgical implantation of the DBS system, PMDS patients will spend seven days in the neurology ward. PMDS patients will also be brought into the ward if they have a serious infection or DBS hardware-related complication, or if they enter into *status dystonicus* – these patients require medications and close supervision from a neurologist or other medical staff (ward nurses). Therapists will also visit ward patients, perhaps to provide advice to family members and ward staff on how to physically handle the patient, or to conduct a quick assess-ment (e.g. the physiotherapist may need to screen for musculoskeletal contractures while the patient is heavily medicated). In regard to such activities, the 'flexible' configuration of the ward does appear to facilitate the PMDS' multidisciplinary approach. However, there are other fea-tures of the built environment that appear to have a more significant role in configuring the multidisciplinary work of the team.

The Greater Hospital

Most of the PMDS activities with patients and families take place in other parts of the hospital. Many of the assessment activities that therapists conduct with patients both before and after implantation of the DBS system require a degree of privacy or a large area of space with a specific set of resources.

As a 'multidisciplinary' building, the hospital was designed so that it contains a diversity of spaces and resources. Each ward contains a play area with toys and small desks and chairs for inpatients and their siblings to play and interact with one another. Each ward also contains a kitchen area adjacent to the play area. Here, the families of inpatients can prepare food and hot drinks while watching children play. These areas of the wards mimic a generic domestic kitchen-living area space. The hospital also contains a school with full-time teachers. This is located in a large mezzanine area of the hospital. It too resembles a 'generic' school environment: student artwork, alphabet and multiplication-table posters are littered across the walls, and it contains desks and chairs of various sizes. The clinical director explains that these features reflect a belief that the hospital is not treating individual children, it is 'dealing with families':

> You always have to remember that a child is part of a family and anything that happens to the child has ramifications for siblings and parents . . . The decisions we made when we were setting up the school is that actually if we're really going to help the children who are in hospital, we need to think holistically about the whole family and we need to think about the siblings who can sometimes almost be more traumatised by what's gone on than the child who's sick . . . The hospital school will take siblings as well as the children so they don't miss a week's schooling. They are doing something that's actually quite fun and keeping them absorbed and engaged. (Clinical Director, interview)

We can see these features of the hospital as reflecting what could be called a clinical concern for children's wider social context. Features of normal domestic and schooling scenes have been recreated within the hospital with the intention of minimising disruptions to the stable family unit.

Importantly for the PMDS team, these areas can be adapted without too much difficulty into the spaces required for conducting pre- and post-surgical assessments. For example (as we will explore in greater depth in Chapter 7), one assessment involves observing and quantifying the patient's ability to undertake tasks of daily living such as brushing teeth, preparing cereal, or washing dishes. The kitchenettes – although not specifically designed as places of assessment – contain the appropriate arrangement of facilities for this assessment with some patients. Other assessments require the child to perform drawing and writing tasks. These can be conducted in the play areas or the school, both of which contain desks and chairs of appropriate sizes. The hospital also contains a large gymnasium. This large room contains a variety of resources: inflatable balls, toys, adjustable desks and benches, padded floor mats of various sizes, and plinths and hoists for moving severely disabled patients. The gym was designed to be an area for providing rehabilitative assistance to children, but for the PMDS, the gym also provides a suitable location for assessing gross motor function (as we will see in Chapter 5), and for other assessments which require open space such as the assessment of impairment. Here, plinths and hoists are essential for placing severely disabled patients into the body positions necessary for conducting DBS-related assessments, and the available toys can be used to entice younger children to participate.

The creation of domestic and educational settings and the provision of a gymnasium afford a variety of clinical activities. PMDS team members have adapted settings into various discipline-specific spaces for conducting activities; activities that (as we shall see in more detail in subsequent chapters) have been fundamental to the adoption of DBS as a therapy for managing dystonia in children and young people. The greater hospital, along with the open-plan office and ward spaces, constitute what could be called a materialisation of a pro-multidisciplinary and pro-teamwork ideology. A particular socio-political trend in health service provision that encourages multidisciplinary teamwork (and which aligns with the neurodevelopmental model of childhood development) has been encoded within the physical structure of the hospital. Obviously, this does not induce multidisciplinary teamwork in a deterministic fashion, but it no doubt enables and prompts individuals to participate in multidisciplinary activities: it is, as Prior has argued in regard to space in general, an active ingredient in producing and reproducing particular social practices (Prior 1988, 91).

The importance of the built environment for the team's activities was made explicit by team members during team meeting discussions concerning the moving of some of their activities to a new location. A steady rise in the number of patients for the hospital as a whole is placing increasing pressure on available resources. In order to relieve some of this pressure, managers within the Trust were encouraging the PMDS to move many of their activities with patients to newly refurbished spaces within a new neurology centre located within the neighbouring adult hospital. This move was the focus of much discussion during team meetings, and the therapists in particular felt that it would hinder their ability to work. The OT, for example, felt that the new rooms were inadequately resourced to safely conduct assessments with children: they lacked hoists and benches for physically managing patients, and they lacked phones which the therapists used to coordinate activities. There was also concern about the additional distance that patient and families would have to traverse. The OT claimed, for example, that it would almost double the time required to conduct some of the assessments.

These discussions about the move also revealed the importance of the team's administrator in coordinating the team as a whole. If the move to the new centre was to take place, some duties currently undertaken by the PMDS administrator (which we explore in the next section) would need to be delegated to the new centre's administrator. Team members felt that the new administrator would lack the knowledge of the team's 'idiosyncrasies' and their unique patient cohort – knowledge that is required to coordinate room bookings with team members' schedules and a patient's clinical needs.[3] When it was suggested by Dr Martin that the new administrator could be informed of the 'blocks of time' that would be needed in advance, the therapists responded that this would not provide the flexibility required when dealing with such a variable cohort of patients: it is often difficult to predict how long an assessment

---

[3] This is an example of what Hardy and colleagues (Hardy and Turrell 1992) referred to as a procedural barrier to interprofessional practice: the planning and scheduling procedures of the new centre are mismatched with the workings of the team.

with a patient will take. Such discussions provide a glimpse of the types of pressures that can disrupt the activities of the team. Despite the suitability of the children's hospital for multidisciplinary teamwork, resource constraints and administrative complications threaten the PMDS' ability to fully utilise it.

Managing the PMDS Schedule: Crafting a Programme of Actions

The second 'binding' strategy of the PMDS that we will explore is the team's use of a coordinating schedule (Fig. 4.2). Much like any schedule or diary it lists the times and locations of upcoming PMDS activities. The arrival and departure times of patients, the clinical activities of each team member, and the various locations where clinical activities will take place, can all be represented on the schedule. Any team member can consult the schedule to see which patient they are meeting, which other team members will be accompanying them, what activity will be taking place, and what room has been booked for them. During my time with the team, I noted that it was often displayed during team meetings via PowerPoint projection. It was consulted, debated, and altered, and it was clear from discussions that team members placed a great deal of importance on the schedule.

The schedule can be seen as what Actor-Network theorists have referred to as a *programme of actions* (Latour 1992; Akrich and Latour 1992). This is an entity that functions as a set of instructions or a script, detailing how other entities (human and non-human) should act and interact at specific times and locations so that an specific output can be generated. An obvious example of this are software programmes which stipulate how diverse components of a machine should interact in order to complete a specified set of tasks. A successful programme of actions enables the various unique capacities of a range of diverse entities to be coordinated and orientated towards the achievement of a set of collective goals. This is the intention of the PMDS schedule: it allocates the actions and interactions required from a heterogenous group of

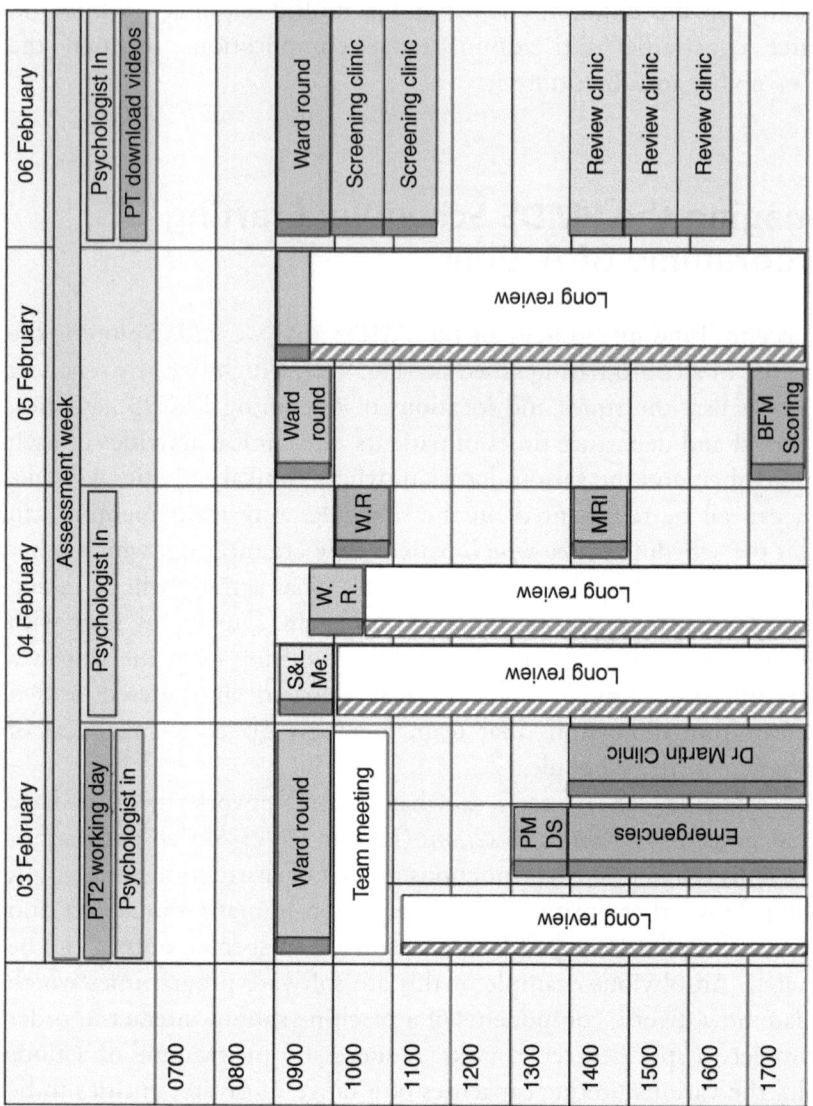

Fig. 4.2 A recreation of the PMDS weekly schedule (anonymised)

professionals so that each new referral can be adequately screened, and so that each DBS patient can be suitably assessed and managed. By doing this, the schedule provides a degree of structure and routine to PMDS activities. As team members consult the schedule week after week, a diverse group of professionals, patients, and their family members are enrolled in and perform the institutional routines that characterise the PMDS, while (ideally) achieving the aims of the team as a whole. Importantly, as May and Finch have noted (May and Finch 2009), such coordinated allocation is often an important aspect of successful technology adoption: it helps to ensure that individual actors are aligned towards the operationalisation and routinisation of the technology. Hence, I suggest that the PMDS schedule – along with the knowledge required to construct it and utilise it – is an important part of the PMDS proto-platform. The construction of the schedule is also an example of what STS theorists have referred to as *invisible work* (Star and Strauss 1999): work which is absolutely necessary for the functioning of the team, but which would tend to be ignored in conventional accounts of medical innovation.

Creating and updating the schedule is an 'ordering' process, in which a diverse array of elements must somehow be coordinated into a single programme of actions. It is undertaken by the team's administrator and the nurse together, usually several times a month, during what they refer to as a 'scheduling session'. During an interview, the administrator described the session as her 'biggest challenge': patients with very different requirements must somehow be booked in to see clinicians with diverse, disciplinary-specific tasks to conduct, some of which require several hours to complete. Patients cannot, then, simply be booked in pre-defined 30-minute time slots. In this section we see how the administrator, with assistance from the nurse, creates and maintains this programme of actions. Various complexities must be anticipated, coordinated, and ordered by the nurse and administrator in order to create a coherent schedule. This activity, I suggest, aligns with what Carmel (2013) has referred to as *craftwork*, which, Carmel argues, is characterised by two features: the application of different knowledges and practical, reflexive interaction with the object being crafted.

Knowledges and Reflexive Interaction

Scheduling sessions usually take place at the nurse's desk in the middle of the PMDS team seating area of the open-plan office. The scheduling task is carried out with a computer and the administrator can easily move between several patient databases and the 'Calendar' feature of a common software programme that provides the schedule template. In creating the schedule, the nurse and administrator must balance several considerations: routines of the institution within which they are based; the capacities of each team member, financial considerations, and the patients' clinical and personal details. In managing each of these, the nurse and administrator draw on and coordinate several knowledges, thus crafting a programme of actions.

A 'blank' schedule already contains a great deal of information. The starting template of the form reflects institutional routines of the greater hospital: working days and working hours are clearly delineated (Monday to Friday, 8am until 6pm), and all PMDS activities must be allotted in this conventional and institutionally defined temporal structure. Secondly, various activities have already been entered within this pre-defined temporal structure, each indicated by a coloured 'block'. These are the standard clinical activities that take place every week largely regardless of 'contingencies'. These also reflect various institutional pressures. For example, as consultants employed by the NHS Trust, both neurologists are obliged to conduct a daily round of the neurology ward (9am to 10am), and Dr Martin is obliged to run a neurology screening clinic on Monday afternoons. Other routine weekly activities include the PMDS screening clinics on Thursday and Friday mornings and the neurosurgeon's visit on Friday afternoon. Importantly, these blocks of activity have been colour coded using a system devised by the administrator and nurse. The colour of the block indicates the impact it will have on the scheduling of additional activities, and by implication, the importance of the individual involved:

> Red is important meetings or activity that is not flexible, that is important for the whole team to take into consideration when scheduling patients.

Dr Martin's [screening] clinic on a Monday afternoon means that he will not be available to see PMDS patients at that time ... Green means that that particular member of staff is not available for that period of time but they're not like a doctor. These are [things] that don't necessarily affect the time-tabling of patients but are useful for others to know. (Administrator, interview)

At the beginning of the scheduling session the diary is already loaded with meaning: it reflects particular institutional patterns and it contains colour codes that indicate the importance and availability of various team members.

As they craft (Carmel 2013) the schedule, the nurse and administrator mindfully respond to these meaningful elements, adding additional elements in the processes. The first blocks of information that the nurse and administrator add to the Calendar spreadsheet relates to the availability of team members. As Akrich and Latour note (Akrich and Latour 1992), crafting a workable programme of actions requires knowledge of the capacities of each of the agents that will perform it: the nurse and administrator require an understanding of the professional role of each team member and an awareness of each team member's availability. As the administrator states, each member of the team has their own individual commitments that need to be considered before patients can be scheduled:

The biggest challenge for me, with it being a multidisciplinary team, is diary scheduling. You've got eight professionals who take leave, go to conferences, have their own research and other things that they do.... (Administrator, interview)

One of the first scheduling tasks for the nurse and administrator, then, is to note the availability of each team member using the colour coding system. The open plan office arrangement means that they can easily check details with team members.

Once these details have been added, the nurse and administrator can book patients for that week. The first set of patients to be booked in are new referrals. These will be booked into the routine weekly PMDS screening clinics on Thursdays and Fridays that are used to identify potential DBS candidates from the many referrals received by the team. Revenue generation determines how many patients and clinicians will be allotted to these clinics. While the revenue from these clinics will only just cover costs, it does allow them bring in patients that will guarantee revenue for the service as a whole. As one of the therapists pointed out during a discussion:

> For it to be financially viable [with that tariff], we need to see two new patients in a slot. We won't be making any money [from that specific screening clinic], but it enables us to weed out some of the patients and initiate needed actions (Physiotherapist1).

The tariff for this activity must cover the cost of a neurologist who is required to assess, among other things, the patient's medicine regime, and a therapist (either an occupational therapist or physiotherapist) who is required to assess the patient's functional concerns. Consequently, when booking new referrals the administrator and nurse ensure that:

> Dr Martin sees two new patients every Thursday morning and [Neurologist 2] sees two new patients on Friday mornings. One therapist is also present (Administrator).

Two new referrals will be allotted into these clinics on each day along with a neurologist and one of the available therapists. These details are entered into the diary and the clinic is coded 'purple' to indicate 'screening'. Importantly, only new referrals that have a 'letter of guarantee' from their regional commissioning group (which were then Primary Care Trusts) stating their willingness to fund the DBS service will be booked in to these clinics.

Next, the nurse and administrator must book patients requiring pre- and post-surgical assessments. Because these assessments can take a great deal of time and require many team members, they can be difficult to align with the other elements on the schedule:

> Trying to fit a child in for a review [post-surgical assessment] at a time when all of the required people are available within a set timeframe, and

when you've got several hundred kids on our service now, many of whom are now being reviewed – it can be a struggle. But that's something that we just have to come to terms with. It's not going to get any easier. (Administrator)

The nurse and administrator consult a database which contains all PMDS patients and indicates the date of their last visit to the service. From this the nurse and administrator can identify those that are due for an assessment. These names can then be 'cut & pasted' into the schedule if an appropriate space with the necessary time and staff members can be found. Importantly, however, the nurse and administrator must take into account other considerations when doing this. These are best illustrated from the following discussion that took place between the administrator and nurse as they

Adm: Friday's are [the patient's] day off college, so that works well for him. They come in from [county] so mid-morning onwards would be ideal. It is a three-hour round trip.
Nurse: It will have to be 11am.
Adm: That will do. It is not too early for them.

The nurse and administrator consider various details about the family and the patient, such as schooling commitments and their living location. Those families that have to travel some distance will be booked into Trust-provided accommodation, the availability of which will also affect the scheduling of patients. Obviously, then, the scheduling process requires the nurse and administrator to become sufficiently familiar with patients and aware of such details. Indeed, in their wider roles within the PMDS, both staff members (along with other members of the team) do acquire such knowledge of families. It is, the administrator states, 'the nature of the service':

I'm the first point of contact for the patients . . . some of them end up talking to me quite a lot if they make contact with the team quite a lot. . . . I do my best to go down and meet as many of the patients as I can because obviously that is the nature of the service. (Administrator)

Clinical knowledge is also needed when crafting a workable programme of actions, and it is for this reason that the nurse is involved in the process. Before pre- and post-surgical assessments can be placed within the schedule, the nurse needs to ensure that the clinical needs of the patient can be met by the available staff. As she explains:

> Our administrator, she doesn't understand all the illnesses, whereas I know their clinical details, what they're coming in for, what clinical need they have. [That's why] I help arrange the whole diary and everyone's life within the service. (Nurse, interview)

The assessment is 'booked' once a time period within the schedule has been found that is manageable for families and corresponds with the availability of the required staff members. Here, then, the nurse and administrator take note of the red- and green-coloured 'blocks' within the schedule to determine which staff members are available. The assessments are coloured white on the schedule to indicate that many of the staff members will be involved (text is added to note which specific team members are involved), and other colour tags are added to indicate whether or not booking details have been sent to the family.

Once the pre- and post-surgical assessments have been added the schedule is almost full. The remaining free time can be used by members of the team to complete paper work, but often emergency bookings will occupy these times. Emergency bookings might be made for patients with infections, *status dystonicus*, or with suspected hardware failures, or for patients who have inexplicably deteriorated.

By examining the schedule crafting process we can see how the itinerary of activities of the PMDS is shaped by a broad range of concerns and not just the clinical needs of the patient. Creating the schedule requires the administrator and nurse to draw on their knowledge of the capacities and availability of each team member, personal knowledge of families, and a clinical understanding of the patient's requirements. These 'complexities' must be coordinated with an institutionally defined timeframe, institutional obligations, and funding pressures. It is via this process of crafting that the administrator and nurse create a PMDS programme of actions: a schedule that describes and prescribes the actions required from each team member so that the overall goals of the PMDS team can be achieved. Importantly, it also functions to integrate the unique, potentially disrupting PMDS team activities with the workflows of the hospital. We can say that, as team members adhere to and enact the schedule, the PMDS as a multi-disciplinary service is 'performed into being' in a manner that is accommodated within the wider institution. This is not to say, of course, that the product of the session is a 'completed' schedule that is diligently adhered to by all members of the team. The schedule is constantly subject to alteration as families cancel meetings and as emergencies arise, and team members inevitably engage in what Akrich and Latour (1992) refer to as *antiprogrammes*[4] such as taking sick leave. Sometimes the nurse and administrator have simply not been able to anticipate what a patient requires from the PMDS. As the administrator explains:

> Sometimes we make the right decisions and sometimes we don't. And [team members] email us to say, 'Why have you put this patient in here? I need to see them at that review and I'm on annual leave.' If you've just spent three hours going through the diary, making all these changes, and then you get an email saying, 'Well that's not going to work, that's not going to work,' it can be a bit, you know, it makes you feel a bit deflated sometimes, because then we have to go back and do it again. But it's just all par for the course really. (Administrator)

The Team Meeting: A Forum for Group Decision-Making

The third 'binding' feature of the PMDS that we explore here is the weekly team meetings. These are scheduled for every Monday between 9:30am and 10:30am, although it was not unusual for a meeting to take up to two hours (to the noticeable frustration of team members). The

---

[4] Individual actions that do not correspond to those of the collective.

team meeting has several purposes. It provides a forum to: discuss current clinical cases and decide upon an appropriate course of action for each case; assess which new referrals should be allocated a slot in the PMDS screening clinic; and to address any concerns that individual team members may have about the service as a whole. In effect, the team meetings represent a regular, structured time for multidisciplinary decision-making, and for talking about and resolving any tensions that threaten to upset the functioning of the team as a whole – both of which, commentators have argued, are vital for effective multidisciplinary service provision (West and Slater 1996; Housley 2003). Such practices entail a *communal appraisal* of DBS therapy. Communal appraisal, as May and colleagues argue in their discussion of Normalisation Process Theory (May and Finch 2009), is an important element in the implementation and routinisation of new healthcare technologies. The meeting enables team members confer on the apparent effectiveness of DBS with particular patients, consolidate their general support of the technique, and they can subsequently reconfigure service activities if needed. In this section, we draw on extracts of team meeting discussions to demonstrate some of these points, and in doing so, we also see how the multidisciplinary arrangement of the team has implications for the way in which patients and their illnesses are rendered intelligible.

Authority and the Coordination of Key Documents

PMDS team meetings are informal. There is no chairperson, and to an outsider such as myself, the meetings may initially appear to be rather disorderly. Structure was provided by the agenda which was projected on a screen on one side of the meeting room (see Fig. 4.3), and as the team worked their way through the agenda items, one team member – often the administrator or one of therapists – would add comments and actions to each item, effectively converting it into minutes of the meeting. In the absence of a chairperson, no one individual kept the team moving through the agenda points or coordinated the utterances of the other members, and it was not uncommon for several team members to be talking at the same time or for discussions to jump back and forth

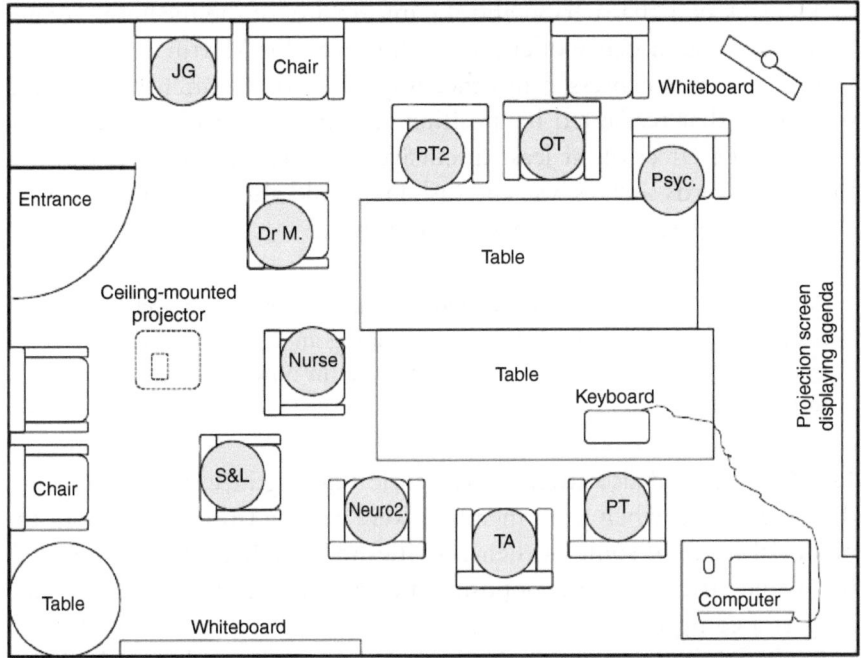

Fig. 4.3 Schematic of the usual team meeting seating arrangement

from one action point to another. A direct, upfront style of communication was a common characteristic of team meetings and several
members of the team, particularly the occupational therapist and the
senior physiotherapist, were outspoken. This no doubt reflected what
several team members described – in a positive tone – as the 'strong
personalities' within the team, and it meant that exchanges appeared to
be rather terse. Any tension, however, was constantly diffused with a
liberal use of humour,[5] and I was explicitly reminded that 'team members get on very well, even if it does not always seem so' (Therapy
assistant).

---

[5] In one meeting, for example, the nurse told the team that one of their older, more functional
inpatients had managed to sneak his girlfriend into the neurology ward for the night, much to the
chagrin of the ward nurses.

The lack of a chairperson and the informality of the meetings are not indicative of an absence of authority, however. Dr Martin's authority as team leader was expressed in other ways. He would often put forward utterances that reasserted his position as team leader; utterances which also had the effect of, at least discursively, 'unifying' the team through the use of words such as 'we' and 'us'. He would also prompt the team as a whole to reflect on the service. For example:

> We need to be very clear about what research means. Research is about audits . . . When we collect data, we are doing an audit and that provides information that enables us to make changes in how we manage patients in the future.

Dr Martin would also attempt ease tension during meetings by playing a diplomatic role: when team members were voicing frustrations about the potential move of some of their activities to the adjacent adult hospital, he would articulate the opposing perspective of 'management' and suggest a compromise.

Dr Martin's actions during team meetings can be understood reifying the team itself, encouraging the team to accommodate with wider institutional pressures, and prompting team members to engage in communal appraisal of, for example, 'how we manage patients in the future'. Another important feature of team meetings is that it provides an opportunity to discuss and manage team documents such as the questionnaires that are sent out to new referrals and the multidisciplinary team (MDT) report that is completed for each patient. The latter contains a summary of the patient and a set of recommendations from each PMDS clinician, and they are usually sent to the child's local community services. A recurrent topic of discussion was the slow rate at which MDT reports were being completed, and the senior physiotherapist was particularly vocal in encouraging other team members to complete their contributions to reports quickly. Such documents are a small component of the network of heterogeneous entities (Latour 1992) that constitute any modern organisation, and within the PMDS, then, the team meetings were an important forum for reaffirming their importance and reminding team members of their obligations to complete and circulate them.

A Space for Group Decision-Making

The main purpose of team meetings, however, is to provide a space for discussing clinical cases and engaging in clinical decision-making. As Centellas and colleagues have argued (Centellas et al. 2014), team meetings can provide an opportunity for an 'interactive process that facilitates interdisciplinary collaboration by aligning different disciplinary identities' around particular objectives 'without necessarily blurring or softening disciplinary identities or local meanings' (Centellas et al. 2014, 313). During PMDS team meetings, various types of specialist knowledge and expertise are articulated and filtered into the decision-making process. As this happens, disciplinary boundaries are demarcated while 'multidisciplinarity' is enacted, and as we will now see, a 'composite' account of the patient and their illness is produced.

During team discussions team members would of course articulate knowledge that reflected their particular area of expertise. We can see in the following extract, for example, that the physiotherapist (PT1) refers to 'movement control' and 'functionality' reflecting her interest in gross motor function. Here the team is discussing Emma, who has a complex and severe movement disorder, and who has had the DBS system implanted for several years:

PT1:	She has been deteriorating. Her selective movement control is certainly getting worse.
Dr M:	Has she grown? Because there is a mechanical disadvantage that can arise when you grow but the muscle can't compensate with the new mass . . .
PT1:	I was getting her to lie on her tummy and lift her leg – it seems to me that the signal is just not getting through. It is not simply about weakness. Qualitatively, her dystonia is getting worse and functionally she is deteriorating.
Neuro:	We had to adjust her stimulation parameters. She wasn't responding well to higher stimulation doses.
Dr M:	Yes, with higher stimulation there is the possibility of inducing Parkinsonian symptoms.

The neurologist Dr Martin (Dr M) employs a different lexicon while discussing Emma: terms like 'stimulation', 'stimulation parameters', and 'voltage' were often used by the neurologists when contributing to discussions. Typically, they would also report on patients' medication regimes, brain imaging results, and in some cases provide a summary of surgical procedures – all of which align with a broadly biomedical model-based understanding of patients. The following is a good example of this:

> *Dr M*:   There is a problem with the [DBS] connection. His left arm became very dystonic and I have admitted him and I'm glad that I did that because he went into status dystonicus. [The neurosurgeon] came over and we went over the images and it seems the extension wire and the electrode connection has become adrift. When I changed the stimulation parameters I managed to get him out of his status dystonicus, and I have put him on a tiny bit of [medication].

Not surprisingly the psychologist's (Psyc) contribution to discussions tended to include utterances pertaining to mood or cognition. Here, she uses the terms 'self-esteem', 'behavioural issues', and 'suicidal ideation', while the team discusses Jimmy, a DBS candidate who is uncertain about whether or not to proceed with DBS.

> *Psyc*:   Well, he is an interesting one really. He has personality and behavioural issues. I think he is lonely and he has poor self-esteem. He has never had suicidal ideation but he definitely wishes he wasn't alive, which is not the same as ideation. His family says that what happened to him was an accident, but it may have been an extreme move to seek attention . . . He and his family are very unsure about DBS.

Similarly, the contributions of other team members reflected their particular specialty area. For example, the speech and language therapist: 'When I last saw her I noted that she speaks very quickly. Once she gets speaking she has trouble stopping', and: 'She has some freezing when eating, but there is no danger with her swallowing . . .' The nurse's contributions involved references to medications, pain, infection, and infection-control practices. Importantly, as clinical cases are discussed and team members make their particular contributions, a composite picture of the patient emerges: the patient is understood in terms of their gross motor function, stimulation parameters, mood and self-esteem, feeding capacity, and functional ability. Team members discursively enact a patient who is more than just a biomedical entity.

In addition to contributing their own discipline-specific understandings to discussions, team members would also draw upon an awareness of patients' families and wider social circumstances. All members of the team appeared to display such awareness. No doubt this reflected a familiarity with families that developed during the considerable amount of time they spend with the PMDS (particularly during pre- and post-surgical assessments). In some discussions, for instance, team members vocalised the opinions or worries of a family or family member. In the following extract the team is discussing how to move forward with Jimmy, the patient mentioned in the earlier extract:

Psyc: He and his family are very unsure about DBS.

PT: Clinically, you want to offer him DBS. He is the sort of case that responds well.

Dr M: I have met him! I think he would make a great candidate for DBS.

OT: His family is reluctant because we can't tell them exactly how he will respond.

Psyc: I think we should put him in touch with another family [that have had DBS]. They still won't be able to get the sort of certainty they want though.

Neuro: Let's put them in touch with Colin. That will give them a broad range of what to expect.

Jimmy, as the Dr Martin and the physiotherapist note, is a good candidate for DBS, but in light of the team's understanding of the family's perspective, they decide to put them in touch with another family that can offer advice. With some clinical cases team members would draw on their understandings of a patient's relationship with siblings and parents. In the following extract, for example, the team is discussing Ioannis, a five-year-old who appears to be highly agitated:

> *Dr M*:   We need [the psychologist] to have a discussion with the family. The family is in distress – let me tell you what has been going on. Ioannis has been waking up early in the morning, screaming... and then in the evenings they are having trouble getting him into the bath. I think there is a large behavioural component to what is going on: I think he is intensely jealous of his younger brother who is running around, playing... It is difficult for the family but I told them we can't use medications inappropriately... I witnessed his interactions with his family. It is clear that it is one of those classic situations where he is spoiled on the one hand but there is a huge emotional entanglement on the other.

Indeed, such comments, in which the well-being of a family are taken into account, were common in team meeting discussions. Team members would also refer to families' access to community services, and as the following extract illustrates, the patient's performance at school and their relationship with peers:

> *Dr M*:   [The dystonia in] his legs are much better. I put him on double [stimulation] contacts and I reduced the right lead voltage to avoid dyskinesia in his face.
>
> *S&L*:   His speech was good.
>
> *Psyc*:   And he seemed cheerful when he was here.
>
> *Dr M*:   On that point, it would be interesting to see whether his mood problems are because now, even though his dystonia has improved with DBS, he still can't keep up with his peers at school. This could be making him frustrated and stressed. Prior to DBS, he could always attribute his inability to keep up with dystonia. Perhaps now that he has DBS, he and others are placing too much expectation on him. This is in contrast to the changes in mood resulting directly from stimulation itself.
>
> *Psyc*:   His dad certainly does not think it is directly a result of the stimulation... you would think the school would be understanding and make allowances for him

Thus, references to 'schooling', 'family dynamics', 'community services', and so on were deployed by team members to understand the patient and thus inform decision-making. Within team meetings, then, patients were enacted not just in terms of gross motor function, pain, medication regimes, stimulation parameters, functional abilities, and so on, but also as individuals within a wider network that includes families, schooling and peers, and community services. Unsurprisingly, discussions and decision-making that involved these frames of reference tended to draw on prevalent social norms. For example, team members make judgements on whether a patient has an 'overly dependent' relationship with a parent, leading to 'emotional entanglement' (as with the case of Ioanis). Schooling and interaction with peers are talked about as things that should be encouraged, and it was common for team members to make assessments and recommendations based on their perceptions of age-appropriate behaviour. A good example of this is the case of Carl, a 16-year-old patient who is about to have the DBS system implanted:

PT1:   I really think Carl should have a non-rechargeable battery. I was really glad when [the psychologist] agreed with me! Dr Martin has been pushing for him to have a rechargeable battery. A non-rechargeable battery would be much better for Carl – he is at that age where he will want to go out and do things. He can't be expected to routinely be at home to recharge his battery.

The Broad Remit of the PMDS Clinical Gaze, and Some Elements of a Proto-Platform

As we have seen in the aforementioned examples, team meetings provided a regular structured time for addressing tensions that could threaten the functioning of the team, and they provided a forum for multidisciplinary discussion and decision-making. As various team members contribute to a discussion, a clinical case is rendered intelligible in terms of a wide variety of elements, from brain imaging and stimulation parameters to relationships with parents and performance in school. Inevitably these 'patient-centred' discussions involved drawing on particular understandings of 'normal' childhood and teenage development.

In *The Birth of the Clinic* (1963/2003) Foucault describes the emergence of what he referred to as the medical gaze of modern medicine. This gaze, he argued, sought to delineate and define the body in terms of its concrete shape and form, and its emergence signalled the birth of the biomedical model of disease and the body in which a patient's social circumstances are largely elided or considered irrelevant to clinical decision-making (Fox 2012, 143). Within the PMDS team meetings, something quite different to Foucault's medical gaze is in operation. Biomedical considerations are certainly mentioned within team meetings and such considerations influence a great deal of PMDS clinical decision-making, but social aspects are also foregrounded. In addition to talk about brain images, medication regimes, and surgical considerations, PMDS team members devote considerable discussion to non-biomedical considerations: schooling, relations with peers, family dynamics, a patient's mood, and so on. We can say, then, that the PMDS casts *a broad clinical gaze* over their patients. I define this as a gaze that *extends from the inner concrete shapes and forms of the brain, to the subjective thoughts and emotions of their patients, and to the internal dynamics of domestic life.* Like the medical gaze described by Foucault, there are normative consequences for those subjected to the broad clinical gaze of the PMDS. Patients are, in effect, subject to a broad surveillance where they are inevitably compared to various norms and where clinical action is undertaken with the intention of upholding such norms – the decision to implant a non-rechargeable neurostimulator in Carl so that he can 'go out and do things' like a normal teenager is an example of this. Bourret (2005) has argued that particular forms of clinical work – and particular ways of perceiving disease – emerge from specific collaborative arrangements. He illustrated how the development of new forms of molecular clinical practices in breast cancer genetics was predicated on the emergence of MDTs involving biologists and specific clinical specialists. These novel clinical practices (which constitute what could be called a biomolecular gaze) emerged from the pooling of the heterogeneous resources from these disciplines and the subsequent blurring of disciplinary boundaries (Bourret 2005, 57). Similarly, we can see here how a 'comprehensive' clinical gaze has emerged from a specific form of multidisciplinarity collaboration.

What we have seen in this chapter is that this type of collaboration is an achievement: it requires a great deal of work. Teams such as the PMDS (or indeed any organisation) do not simply exist in the 'order of things'. Rather, they are brought into being and maintained through various practices of socio-technical organising (Alcadipani and Hassard 2010). I have explored several of these practices: the careful arrangement of the built environment which was purposefully designed to encourage multidisciplinarity; the scheduling session which is used to enrol each individual team members into an collective itinerary of actions; and the team meetings itself, which provides a forum for reminding individuals to contribute to and circulate the various team documents that are necessary for effective functioning of the team. There are, of course, numerous other practices that have the effect of maintaining and unifying the multidisciplinary PMDS team. For example, team members regularly socialise together, which no doubt reinforces a sense of friendship and trust. The team also maintains a webpage which delineates the team and presents it to the public as a 'comprehensive service'. And importantly, tools and protocols used within the PMDS can have a 'unifying' influence: some assessments, for example, can be carried out by several members of the team, and some may require several team members to work together – the goal setting session which we explore in Chapter 6 is a good example of this. The PMDS can be seen as the ongoing consequence of these everyday practices of ordering and unifying, and it is upon these types of practices that the broad clinical gaze is predicated.

Foucault argued that the modern medical gaze was encoded in and perpetuated by various institutional arrangements: it was not simply a mode of perception and reasoning that existed in the minds and practices of observers and doctors (Foucault 2003, 109). Similarly, in this chapter I have indicated that various institutional and contextual factors have meant that the barriers to multidisciplinarity are perhaps not as imposing for the PMDS as those encountered in other areas of health and social service provision. The prevalent neurodevelopmental model of childhood development and an accompanying pro-multidisciplinarity discourse have, while influencing the PMDS structure itself, also informed various aspects of their environment, such as the payment structure which has reduced the financial boundaries for teams such as the PMDS, and of course the structure of the hospital itself, which has permitted particular forms of multidisciplinary. Such arrangements provide 'institutional weight' and durability to the broad clinical gaze.

What we have also seen, then, is how such arrangements have collectively shaped the features of an organisational form that has emerged in the process of implementing and routinising DBS within paediatric neurology. I suggest that that the payment system, the spaces within the hospital, and the neurodevelopmental 'way of thinking', along with the PMDS team structure and the team schedule, are elements of a localised *proto-platform*. They represent, in other words, mutually configuring heterogeneous elements of a nascent socio-technical infrastructure; an infrastructure for operationalising DBS as a therapy for managing dystonia in children and young people.

Summary

In this chapter we have explored the challenge of coordinating multidisciplinary teamwork. Multidisciplinary teamwork, or 'interprofessionalism', has been championed as a means of providing comprehensive patient-centred care, and various health policy initiatives have set about encouraging multidisciplinary approaches in health and social care contexts. Yet, multidisciplinary approaches can be difficult to operationalise due to structural and institutional barriers, traditional hierarchies among health professionals, and incommensurable discipline-specific understandings and viewpoints. In this chapter we have closely examined some of the contextual factors and practices that facilitate the PMDS's multidisciplinary approach. These include: the neurodevelopmental paradigm of childhood development which provides a shared knowledge base for all PMDS team members; the architecture of the hospital, which was specifically designed to encourage multidisciplinary teamwork and patient-centred care; the clever use of 'mundane' multidisciplinary technologies such as the team diary – a *programme of actions* (Akrich 1992) which aligns individual team member activities; and regular team meetings, during which the team leader is able to discursively unite the team, and which provides team members with an opportunity to collectively appraise their activities. These various features can be seen as important elements of the PMDS patient-centred proto-platform, and importantly, the broad clinical gaze of the PMDS is predicated on such elements.

In the following chapters we explore how PMDS team members manage three more challenges, and in doing so, we glimpse other important elements of this localised patient-centred proto-platform: embodied knowledge and body–space configurations; tools for managing future-oriented visions; and clinical assessment tools. These too are implicated in perpetuating a broad clinical gaze, and in each chapter I interrogate in more depth just how the broad clinical gaze extracts clinically useful knowledge from the patient.

5

Body Work and Space

The proto-platform elements that we have just explored in the previous chapter are, of course, similar to the types of elements explored by Keating and Cambrosio (2003). In their analysis of the emergence of biomedicine, they too closely examined institutional arrangements, architectural forms, and particular 'ways of thinking', and they explored the implications of platforms on understandings of disease and illness. In this chapter, we focus on platform elements that weren't examined in detail in their analysis, but which would no doubt be implicit in their general argument: the bodies and embodied knowledge of health professionals. As with the other platform elements, embodied knowledge is a means by which a particular way of perceiving and engaging with the world has become embedded in a form that is, to some degree, durable. Obvious examples are bodily habits and tacit knowledge, some of which may distinguish professional groups (Gardner and Williams 2015), some of which may distinguish social classes (as Bourdieu has argued), and some of which may be shared broadly among cultures (as Mauss has argued). Bodies, in other words, have structuring effects: they tend to both reflect and reproduce wider socio-political forms. We can expect then that bodies will have structuring effects on biomedical innovation.

© The Author(s) 2017   **123**
J. Gardner, *Rethinking the Clinical Gaze*, Health, Technology
and Society, DOI 10.1007/978-3-319-53270-7_5

As an illustration, this chapter focuses specifically on the physiotherapists as they use their bodies to do a specific task: identifying appropriate candidates for DBS. Additionally, this chapter argues that such *body work* requires the careful construction of spaces, which subsequently facilitate and prompt particular types of *sensing and acting* within. By doing this I emphasise the importance of the body in innovation work, while also demonstrating that the productive body in biomedicine is not a distinct, well-bounded entity: rather it is momentarily configured by a wider array of (proto-)platform elements. I begin with a more thorough description of the 'adoption challenge' of identifying suitable candidates, and then follow with an overview of some theorisations of the clinician-body in which I delineate two forms of body work: *communicative body work* and *sensorial reflexivity*. As with the previous chapter I then draw on both observation and interview data to closely examine the physiotherapists at work, highlighting instances of both communicative body work and sensorial reflexivity while also noting the importance of the wider space in configuring this work.

The Challenge: Identifying Suitable Candidates for DBS

Well over half the patients who attend the PMDS for DBS have secondary dystonia. In these patients, many of whom have cerebral palsy, dystonic movements co-exist with spasticity, muscle weakness, and contractures, as well as with cognitive and communication difficulties. Of these various manifestations of neurological pathology, it is only dystonia that can be directly managed with DBS. In its 'pure' form the spasms and shaky movement that characterise dystonia easily distinguish it from other motor signs. Spasticity, for example, appears as stiffness and rigidity, the consequence of the excessive contraction of a muscle or muscle group. Yet, the clinical presentation of many PMDS patients, many of whom are the most severe cases in the UK, is complex. In these patients it is not unusual for both dystonia and spasticity to occur in the same regions of the body ('mixed hypertonia') and distinguishing the two becomes clinically difficult (Lebiedowska et al. 2004). Because of this it is often unclear which motor sign is causing the patient the greatest discomfort or has the greatest impact on motor function. This is further complicated by the presence of permanent musculoskeletal abnormalities that can restrict a patient's movement and produce painful posturing.

This presents a challenge for the team. In order to determine if DBS will be of any benefit to a patient, they must be able to distinguish dystonia from other manifestations of neurological pathology, and they must be able to determine to what degree dystonic movements are detrimentally affecting the patient. If, for example, a patient has both spasticity and dystonia but it is predominantly spasticity that is creating difficulties for the patient, then DBS will be of little use. There are no technological solutions to this challenge. MRI and PET may render neurological abnormalities visible, but the images they generate cannot be used to distinguish the various effects of these abnormalities. Current clinical measurement tools are also of limited use as they lack the sensitivity necessary to identify and capture subtle differences (Gordon et al. 2006).

The PMDS' solution to this problem is to rely on the embodied knowledge and tactile skills of the two physiotherapists. During an initial interview with patients and during the subsequent regime of pre-surgical baseline assessments, the physiotherapists (henceforth PTs) conduct an examination of gross motor function and a musculoskeletal screening, both of which are used to identify the various manifestations of neurological pathology and to assess the degree to which dystonia affects the motor functioning of the patient. In order to conduct these examinations the physiotherapists must use their own bodies to extract sought-after clinical information from the bodies of patients. They use their bodies to encourage patients to comply with the examination, and we see that they rely on a carefully honed sense of touch to differentiate dystonia from spasticity.

The Clinician-Body, Space, and Knowledge Production in Medicine

Scholars within medical sociology and the sociology of health and illness have tended to focus on the bodies of patients and those subjected to clinical discourses and practices, rather than the bodies of health professionals. Of those who do explore the clinician-body, we can roughly divide them into two groups. First, there are studies that have explored what could be called the *communicative body*. Here, scholars have analysed the ways in which health professionals use their body to help encourage the compliance of patients within clinical interactions (Heath 2002, 1986; Brown et al. 2011; Maseide 2011). Heath, for example, has explored the ways in which both clinicians and patients use their gaze and body language to indicate recipiency and prompt the other to talk (Heath 1986). Brown and colleagues (Brown et al. 2011) have explored how clinicians use their bodies to convey information to patients, complementing verbal language, within the context of gynae-oncology. For many of the patients included in the study, whether or not they trusted a clinician depended a great deal on the clinician's bodily actions and gestures. Particular forms of body movement and presentation inspired trust and confidence in patients, thus facilitating clinical practice. In a similar vein, Måseide (Maseide 2011) noted how clinicians used their own bodies to instruct patients how to use and move their bodies during examinations. Respiratory physiological exam-inations, for example, require patients to physically interact with tech-nical equipment in a precise and challenging fashion. In order to encourage patients to do this, clinicians would often use bodily gestures in addition to verbal instruction. If this was done successfully, the interaction would generate a textual artefact (such as a note on the patient's medical records) that could subsequently be used to inform further clinical action. Måseide illustrates that communicative body work is essential to ensuring the success of the examination, and suggests that clinical examinations can be seen as 'mutually constitutive processes between various agents, bodies and body modes' (Maseide 2011, 297). These studies illustrate that the body is a key instrument for commu-nicating with patients and ensuring that clinical interactions proceed 'on script'. Importantly, these studies also illustrate that the production of clinical knowledge depends on an interactionally generated, shared understanding, or the achievement of a meaningful clinical space. Communicative body work and verbal communication enable the pro-duction of such a space by endowing elements within it with symbolism and meaning.

The second aspect of corporal work that has been explored (Harris 2011; Schubert 2011) is what Moreira (2004) has referred to as *sensorial reflexivity*. Sensorial reflexivity refers to the sensing-and-acting habits acquired by health professionals via training and clinical experience: the learned perceptual skills and embodied dispositions that may be difficult to verbally articulate and communicate, but are nevertheless a vital component of clinical practice. Drawing on Merleau-Ponty's work, Harris (2011) illustrates that conducting even the most routine, mundane clinical activities such as inserting a cannula requires a tactile competency that can only be acquired through bodily practice, and can easily be disrupted if, say, a new type of cannula is introduced. Several studies of surgeons and surgical training and practice have thoroughly explored this embodied sensing-and-acting (e.g. Prentice 2007; Zemel and Koschmann 2014). Moreira, for example, notes that the surgeons in his study had, via training and clinical experience, carefully attuned their tactile and visual senses to particular phenomena within the fleshy bodies of their patients. Only by learning how to sensorially register these particularities could they then act upon them and thus conduct the surgery. As Prentice argues (Prentice 2007), surgeons are trained to embody schemes of perception and thought.

While the communicative body enables the production of a meaningful clinical space, the 'sensing-and-acting' clinical body is configured as such by a meaningful clinical space. This is cleverly illustrated in Moreira's study (Moreira 2004). The sensing-and-acting abilities of surgeons depend on a wider array of carefully arranged material and semiotic elements: patients are very carefully prepared before surgery so as to present specific body regions to the surgeon, and the operating theatre is precisely arranged so that it enables, constrains, and directs the perceptual abilities of the surgeon towards particular attributes of the patient. Moreira points out, then, that both the surgeon and the patient are the subject of this wider surgical space. The patient's body is configured to present particular sensory affordances, while simultaneously the surgeon is configured as a particular sensing-and-acting surgeon-body. In effect, the surgeon's learned, embodied schemes of perception are 'activated' by a network of material and semiotic elements that constitute the surgical space.

These studies of the communicative body and sensorial reflexivity in surgical practice illustrate the spatially situated nature of the productive clinical body. Body gestures (along with verbal communication) laden a material environment with symbolism and meaning, and a meaning-laden material space configures particular modes of sensing-and-acting within. The relationship between the knowledge-producing clinical body, the patient-body, and clinical space is one of mutual co-constitution. Clinical knowledge emerges from an interaction during which clinical space, the clinician-body, and the patient-body are simultaneously configuring one another. In what follows, I conceptualise the PTs work in this way. The process of identifying dystonia – necessary for identifying suitable candidates for DBS – involves both communicative body work and sensorial reflexivity. Clinical knowledge about a specific patient – usually inscribed in a durable form such as a text – emerges from an interaction in which space, the clinician-body, and the patient-body (or body-part) mutually configure one another.

More specifically, we see that the identification of dystonia requires a carefully constructed material terrain that becomes a productive diagnostic space as it is laden with meaning via talk and communicative body work. This material- and semiotic-constituted diagnostic space enables the generation of what I refer to as momentary affects: temporarily induced patient-bodily phenomena which are registered by (or affect) the senses of the clinician. The clinician then translates this momentary affect into explicit propositions which can be inscribed in notes, medical records, or some other form of durable text, which will later inform decisions about whether or not the patient will be offered DBS.

Musculoskeletal and Gross Motor Function Screening

The physiotherapists use two assessment tools to identify dystonia and to attain some idea of its impact upon the posture and motor function of the patient. These are a manual musculoskeletal screening and a specific means of examining gross motor function called the 'Gross Motor Function Measure' (GMFM).

Musculoskeletal screening can involve any number of standard examinations of muscle strength, joint movement, muscle bulk, and reflexes. The intention of the screening is to evaluate a patient's motor system and locate abnormalities that are causing motor dysfunction. Various regions of a patient's body can be explored one by one and specific sets of examinations can enable the clinicians to decipher if an abnormality is located in a particular part of the peripheral motor system apparatus (such as muscles and neuro-muscular junctions) or the central motor system apparatus (pyramidal system, the basal ganglia, or the cerebellum) (Reeves and Swenson 2008). Conducting many of the examinations requires the physiotherapist to engage in close-contact physical work with the patient: they may be required to manually move a joint or place their body weight upon parts of the patient's body, and in many cases they must be able to 'feel' particular phenomena within the body of the patient. Particular manifestations of neurological pathology, for example, have a signature 'feel', but this is perceptible only if the patient's body is manipulated in a specific fashion. Within the PMDS, the PTs use neuromuscular screening to physically detect and locate spasticity, muscle weakness, contractures, and dystonia. As we will see further on, conducting an effective musculoskeletal screening requires the PTs to possess a carefully honed tactile sensibility.

The GMFM was developed in the 1980s to assess children and young people with cerebral palsy; specifically to obtain detailed information on a child's motor skills, to measure motor skill changes over time, and as a means of assessing the impact of posture aides and braces. In order to conduct the examination the client must attempt a range of gross motor movements, which include rolling, sitting, crawling, kneeling, running, and jumping, in a manner prescribed by a GMFM manual (Russell et al. 1989). Depending on their functional capacity, the client may attempt up to 88 separate tasks, ranging from the very easy such as turning their head while lying on their back, to the more difficult such as jumping forward 30 cm with both feet simultaneously. As this is done, the clinician provides a score for each task using a four-point scale, where zero is 'does not initiate' and three is 'completes'. From these individual task scores an overall score is calculated that is intended to reflect the client's motor function capability (what they can do), rather than their performance (what the child does do) in everyday settings. Clinicians can use the tool to guide clinical decision-making (Tieman et al. 2005): as the assessment is carried out, it provides clinicians with an opportunity to interpret the possible causes of gross motor function impairment, thus providing valuable information that can be used to decide upon an intervention.

It is in this respect – as a guide for intervention planning – that the GMFM is particularly useful for the PMDS. The therapists use it as a means of assessing and quantifying their patients' gross motor function, and along with the musculoskeletal screen, as a means of identifying the possible cause of patients' motor function impairment. As one of the PTs explains, this is not a function of the tool itself; rather, it is a matter of the PT's interpretation:

> The GMFM actually only gives you a number. It doesn't tell you why somebody can't do something. And that's what the professional who's doing it brings to the interpretation . . . The professional carrying out the GMFM interprets as they go about it, from analyzing the quality of movement, and then linking that to the [muscle screening] – if I know a child has weakness around the hips, it wouldn't surprise me that they'd have trouble standing on one leg.

Together, the musculoskeletal screen and the GMFM enable the PMDS physiotherapists to decipher dystonia from other manifestations of neurological pathology, and to produce an overall picture of how each manifestation is impacting upon a patient's gross motor function. As one PT indicates, this information enables the team to assess the suitability of DBS for that patient:

> And it will give us an idea of what they're able to do, what they're not able to do and how the dystonia is preventing them doing something, or whether it's actually weakness, because we need to decipher if they're unable [for example] to stand up, is it the dystonia that's impacting that? Or is it more weakness? And then obviously from our clinical experience, we know that DBS isn't going to directly influence weakness. So it's just kind of deciphering.

In order to successfully conduct both the musculoskeletal screen and the GMFM the physiotherapists use a range of tools and props. As the following section illustrates, these tools and props constitute part of a material terrain that configures specific interactions between the bodies of the physiotherapists and the patient, enabling the generation of sought-after clinical information. In order to explore this in some detail, we follow a specific case: Carl, a 16-year-old patient with secondary dystonia. Carl is accompanied by his mother, and the assessments are conducted by both the PTs.

Constructing a Material Terrain

Before Carl's assessment begins, the physiotherapists go about adjusting and arranging many of the objects within the gymnasium to create a material spatial configuration in which the musculoskeletal examination and GMFM can be conducted. Aspects of this resulting distribution of material objects are prescribed by the GMFM manual and are therefore (ideally) uniform across all GMFM assessments. As the PT explains:

> The GMFM is quite defined about having something at waist height and defines what there is. Various height adjustable benches and table tops and all those kinds of things [are required].

Before beginning the assessment, one of the PTs asks Carl to stand next to the couch so that she can adjust it to his waist height. She also adjusts the bench so that, as Carl sits on it, his feet will be placed squarely on the ground. The second PT pulls two floor mats together and places them nearby to create several metres of padded floor space. Some elements within the gymnasium have been permanently arranged in accordance with the GMFM manual, the most obvious example being the set of guiding floor markings: a series of lines of various lengths and a circle which have been painted on the floor in the middle of the gym. These can be used to help guide the patient perform various GMFM tasks such as walking in a straight line or jumping a specific distance. Rather than having to mark out lines and distances during each examination, the physiotherapist arranged for a set of guiding markings to be painted permanently on the floor of the gym. During an examination a patient can simply be instructed to 'jump from inside the circle to the line'. As the PT states:

> The GMFM are the main lines that are down there . . . I mean the idea [of the markings is] that there's quite set criteria in the test that the patient has to adhere to. For instance, if they walk along the line, they have to have their foot on the line, they can't step off. So having the facility set up ready to go. And also having them predetermined, it means we're all using the same marks each time. Before we had the lines painted on, we often had to push two mats together and measure it every time and get children to walk between the mats . . . But you can imagine there is an error that would potentially affect the child's score. So it's obviously its faster for us and more professional looking, but it also means all the professionals are using exactly the same test criteria to score the children on. So that's the advantage of those.

These guiding floor markings along with the adjusted bench and couch and the padded floor mats create part of a material terrain; a spatial organisation of objects which enables the GMFM and musculoskeletal screen to be conducted and to produce useful clinical information about the patient. As various STS scholars have argued, the material configuration of space is an active, structuring force in the production of knowledge (Gieryn 2002; Henke and Gieryn 2008; Guggenheim 2012). This 'structuring force' is, in part, a consequence of a carefully constructed material terrain that prompts, guides, and channels human activity in such a way that it presents various affordances for knowledge production and action, and it is also a consequence of physical demarcation that shields this activity from noise, pollutants, and intrusion (Gieryn 2002, 48). As we will see, the gymnasium terrain created by the PTs participates in configuring the body of the patient in such a way that his movement disorder is enacted as specific bodily effects, and it participates in configuring the PTs' bodies in such a way that their perceptive skills are rendered sensitive to these effects. For this to occur, however, the patient must be prompted to move and position his body in particular ways, and this requires the achievement of shared understandings. This is achieved through *communicative body work* and verbal communication, which together laden the material terrain with meaning. The material terrain thus becomes a productive diagnostic space.

Communicative Body Work: From Texts and Words to Corporeal Movement

Carl's assessment begins with the GMFM. The role of the PTs is to ensure that he attempts to perform each of the tasks prescribed by the GMFM manual, within the material terrain they have created. While much of this is done via verbal instruction, the PTs also rely on their body as a means of communication. During the GMFM the PTs use both verbal and corporal communication to instruct Carl how he should position and move his body, often in reference to other material objects within the diagnostic space. The following is an example where the more junior of the two PTs (PT2) uses her body to instruct Carl how to perform 'Task 41', which requires moving from 'prone' to '4-point, weight on hands and knees'. (The other PT watches and provides a score for Carl's attempt):

> *PT2:* I want you to lie down on your front, flat on the mat facing me and then get up on all fours. Like this. [She lies face down on the floor flat (prone), and then picks herself up so that she is resting on her knees and hands (4-point)]
>
> *Carl:* [Repeats the task without noticeable difficulty]

The PT embodies a small part of the GMFM text: she performs it, with the intention that Carl will mimic her and do the same. Corporal and verbal communication are used together as a means of complementing one another throughout the assessment, and in the process, bodily movements and words acquire specific meaning. Indeed, verbal utterances are indexical to the PT's body movements and the material and discursive elements that constitute the diagnostic space. An example of this is the PT's instructions on how to perform 'Task 45', which requires the patient to 'crawl reciprocally forward for 1.8m':

> PT2:   The first thing we are going to do is the commando crawl. Get down on that mat for me. You need to pull yourself along the mat and keep your body very low, to the end of the second mat [which is approximately a distance of 1.8m].
> Carl: [Looks puzzled].
> PT2:   Okay, this is what I mean. [She gets down on the floor so that she is resting on her stomach and elbows. She then uses her elbows and knees to propel herself forward.] Remember to keep low.
> Carl: [Carl, without speaking, gets down on the floor and does this same with some difficulty]
> PT2:   Remember to keep low!

Here 'commando crawl' and 'keeping low', descriptions that initially caused confusion, are linked with specific corporal form and movement; they acquire specific referents within the assessment, and in the process, material elements such as the mat acquire meaning. By using her body to complement verbal instructions, the PT is, in effect, participating in the production a meaningful semiotic world: the material terrain is interactionally transformed into a diagnostic space that will subsequently enable the production of diagnostic knowledge.

A consequence of successful communication is that the patient will attempt to perform a series of GMFM-prescribed body postures and movements. Importantly, these postures and movements are prescribed in relation to other material elements within the gymnasium: the patient is prompted to engage in carefully coordinated interactions with the entities that constitute the diagnostic space. Within these interactions, the patient's body is configured so that particular bodily phenomena are framed and amplified in such a way that they can be registered by the perceptual skills of the PTs. Good examples of this are Tasks 54 and 55, which require the patient to stand upright, holding on to a bench with one hand and lifting the right foot off the ground and then the left foot of the ground. PT2 provides instructions while PT1 makes a note of the score:

PT2:   Now, stand next to the couch. Just stand there for 20 seconds as still as you can.

Carl: [Stands next to the couch, but his shaking body makes it difficult for him to keep balance. Several times he has to adjust his feet to keep himself from falling].

PT1:   How much effort does it require to stand still?

Carl: It is hard to stand still. I can't stand still at all.

PT2:   Now, put your hand on the couch, and try and see if you can lift the alternate foot.

Carl: [Does this with the left foot with some effort] . . .

PT2:   One, two, three, well done.

Carl: [Tries with the right foot, but has to place his second hand on the bench to stop himself from falling].

PT1:   Carl, would you mind taking off your t-shirt? I want to see what is happening with your spine.

PT2:   [According to the GMFM manual] you get three goes on each foot. Try standing on it. One, two, three, four, five . . .

Carl: This is the problem, when I try and stand on this [right] foot.

PT2:   Your left foot moves around and throws you off balance, doesn't it.

Carl: Yes.

Here, a GMFM-prescribed body–object interaction involving Carl, the bench, and the flat surface of the floor has generated clinically relevant bodily phenomena: an otherwise obscured motor system abnormality is coaxed by the ensemble to manifest as a flailing foot and a twitching spine. These are momentary affects: temporarily induced patient-bodily phenomena which are registered by (or affect) the senses of the clinician. Mol's (2002) notion of enactment is useful here. Within the GMFM-prescribed ensembles, Carl's motor system abnormalities are enacted as an inability to carry out a prescribed task, or (as we will see further on) as specific tactile sensations. This enactment is collective, as it involves an arrangement of material objects, the compliant patient, and the coordinating and attentive PTs; and it is also momentary, as it lasts only as long as these elements remain within a precisely arranged ensemble.

Aspects of this enacted motor system abnormality will be captured and translated into more durable modes of representation such as a text. During each GMFM-prescribed movement and posture, a PT will take notes and provide a score for Carl's performance. As the following examples show, while the junior PT (PT2) instructs Carl, the other PT (PT1) watches carefully and provides a score for his performance on the GMFM score sheet. The first example, task 72, requires the patient to walk forward ten steps carrying a large object with two hands.

> PT2:   Okay, now I want you to walk holding this ball.
> Carl: [Picks up the large inflatable pink ball which has two small appendages on either side and walks to the end of the guiding lines. Despite appearing a little unsteady, he does this without any obvious difficulty]
> PT1:   Easy peazy! No problems there [She scribbles a 3 on the GMFM score sheet.]

In this case Carl's bodily movements are translated by the PT into a number (3) that is subsequently inscribed on the score sheet adjacent to the specific task number (72). At a later time these scores and the accompanying notes will help inform the team's predictions on how Carl might respond to DBS. Here we see, then, that particular *momentary affects* have been translated into durable clinical inscriptions.

Sensorial Reflexivity: Sensing and Acting

Once the GMFM is completed, Carl and the PTs have a short break before beginning the musculoskeletal screen. By this stage the PTs have a good idea of which particular areas of the motor system are causing difficulties for Carl, and this information is used to decide which areas of his musculoskeletal system will be screened. Carl's shoulders, pelvis, hips, knees, and ankles will be screened for range of movement, muscle weakness, spasticity, and contractures, and this will, ideally, enable the PTs to decipher the presence of dystonic movements.

The assessment takes place on the adjustable couch. Throughout, Carl is instructed to adopt a number of body positions depending on the aspect of the musculoskeletal system that is being screened. Much like the body–object interactions of the GMFM, in each one of these positions the material form of the couch participates in moulding and supporting the patient's body in such a way that clinically relevant bodily phenomena are framed and amplified. And as with the GMFM, the compliance of the patient requires the interactional achievement of shared understanding. However, the musculoskeletal screen requires that PTs themselves become physically involved in the interaction. When testing for range of movement or smoothness of movement (which can be used to detect spasticity), the patient is instructed to remain passive, and the PT will support and move a part of the patient's body in a specific fashion. Standard techniques of musculoskeletal screening, outlined in various user manuals and learned by PTs as part of their training, describe how the PT should use their body in this way. For example, many examinations require the PT to apply 'overpressure' to a particular joint, which involves gently flexing or extending a joint beyond its usual range. While doing this, the PT is instructed to 'be in a comfortable position', use their 'body weight, or the upper trunk to produce the force, rather than the intrinsic muscles of the hand, which can be uncomfortable for the patient', and in order to accurately direct this force, ensure that the their 'forearm is positioned in line with the direction of the force'. All force should be 'applied slowly and smoothly to the end of the available range' (Ryder 2011).

In some cases the patient will take a more active role in creating momentary affects. When testing for muscle weakness, for example, a patient will be instructed to attempt to move a limb while the PT uses her own body weight to create resistance. Here is an example from Carl's assessment. He is seated upright on the couch with his legs dangling off one end, while the PT examines the muscle strength of his quadriceps:

> *PT2*: Carl, I'm going to try and hold your feet down and I want you to try as hard as you can to extend your leg. [She moves to the end of the couch and holds both his feet, one in each hand].
> *Carl*: [Slowly extends his legs, with noticeable effort, while the PT exerts pressure]
> *PT2*: [Talking to PT1]: There is a bit of loss at inner range – I think it was the jerkiness.
> *PT1*: That fits with the movement disorder.

This example also illustrates another purpose of the PT's body within the ensemble: to be momentarily *affected*. As various physical manipulations take place, the PT employs a carefully honed tactile sensibility to assess muscle strength, smoothness of movement, and movement range, and thus help identify motor system abnormalities. Here the body–body–object ensembles are enacting motor system abnormalities as specific bodily sensations. During an interview, the junior PT described how particular abnormalities 'feel' when she is conducting a screen with a patient.

> To see if a child or a young person has spasticity, you bend the knee very quickly. You'll suddenly get like a bony block, it will feel like a bony block but it's not a bony block, but it's like a catch. And then it will release and you will be able to bend it a bit further. Now that is spasticity. Whereas if you have the leg that's straight and you're about to bend it and you struggle to bend it throughout range, but you don't get one of these fast bends where you get a stop, then that's high tone . . . Sometimes you might get a leg and bend it and it's really floppy, so you would call that as low tone. And then other times you'll try and bend a knee and you can't flipping bend it because the quadriceps are kicking in, which is dystonia, it's literally kicking in and stopping you bending [their knee]. It's completely rigid and you can't bend it.

As the PT points out, various motor system abnormalities have their own signature feel. This 'bony block that is not a bony block' sensation of spasticity, which can only be detected when the muscles are passively moved at high velocity, is often described as the 'clasp-knife' response, due to the similar feeling of resistance when closing a folding pocket knife (Burke et al. 1970). Similarly, the difficulty in bending a knee through its range of movement due to high tone (or hypertonia) is been referred to as 'lead-pipe rigidity'. These are, of course, descriptions of temporary, collectively enacted bodily phenomena.

During Carl's assessment, then, the PTs are drawing upon a sensorial understanding of the motor system and its abnormalities. This tactile capability has been acquired during their professional training and has no doubt been honed in clinical practice. They have acquired a body that is tuned to particular sensorial affordances, just as the neurosurgeons described by Moreira (2004) have become sensitive to numerous visual and tactile sensorial differences within an operating theatre. To borrow Latour's (2004) parlance, the PTs possess bodies that have learnt to be affected by and moved by a set of contrasts that many other bodies would fail to register. And as we have seen, the realisation of this sensorial capability requires a specific, material, and semiotic constituted clinical space

As with the GMFM, these temporarily generated bodily phenomena are translated into a more durable form. The following is from an examination which requires Carl to lie on his back, holding his knee in a flexed position while the PT splays his legs:

PT2:   I got some resistance there, and the abductor switched on.
PT1:   I think it is those intermittent jerks. And during and examination of the range of motion of his left knee:
PT2:   Ahh – there is a catch here.

Here, groups of words are used to articulate particular tactile sensations that, as a result, acquire a discursive existence as 'intermittent jerks', a 'catch', or 'some resistance'. These utterances are then inscribed in text: both PTs jot down some of these utterances along with their interpretations ('muscle weakness' 'spasticity', and so on) in handwritten notes, so that by the end of the screening they have produced a document that provides a textual picture of various aspects of Carl's motor system and its abnormalities. There has been, therefore, a series of translations from bodily phenomena to utterance to text, in which the embodied knowledge of the PTs, their ability to physically mould and manipulate the patient, and their honed tactile sensibility, have been essential.

By the end of the GMFM and musculoskeletal screen the PT's have produced sufficient information to provide some sort of diagnosis. Based on their sensorial understanding of Carl's motor system, they declare that he does indeed have dystonic movements. These movements are detectable in regions of his upper body, his left leg, and he has dystonic posturing in his feet, all of which impair his gross motor function. But the main problem for Carl, the PTs inform him, is muscle weakness around the pelvis.

> *PT1:*   Carl, the hip abductors and hip extensors are an issue. The main issues are around your pelvis, it is due to muscle weakness. Maybe we could teach you to do some exercise that could help you there.

This information is also recorded in Carl's medical notes. During a subsequent meeting several PMDS clinicians use these notes to inform a prediction of how DBS may benefit the patient: Carl and his mother are told that it may improve his ability to perform tasks that involve using his upper body, but that it is unlikely that it will directly improve his stability while standing and walking.

Bodies, the PMDS Proto-Platform and the Broad Clinical Gaze

In their analysis of immunophenotyping platforms, Keating and Cambrosio show how platform elements frame and amplify particular bodily phenomena: they entwine these phenomena in particular systems of understanding to produce 'knowledge'; and this knowledge informs subsequent diagnoses and clinical decision-making. The use of flow cytometry to diagnose blood cancers is a good example. The cytometer device casing and internal componentry are arranged in such a way to prompt and channel certain interactions within (a solution contain cells is passed through a laser) while reducing noise and unwanted 'outside' influences. These shielded interactions generate momentary phenomena (a fluorescing white blood cell and the fleeting deflection of the laser) that are registered by the carefully arranged detectors and are then translated into a durable inscription (the deflection pattern). Of course, the successful utilisation of a flow cytometer requires it to be immersed in wider material–semiotic relations, or a 'biomedical platform' (Keating and Cambrosio 2003) involving other machines, reagents, texts, and laboratory technicians with particular cognitive and tacit knowledge.

In a similar fashion the bodies of the physiotherapists play an important role in making the PMDS patients intelligible. They possess a carefully honed set of body skills which enables them to induce patient compliance and subsequently foreground – or frame and amplify – particular phenomena that will subsequently inform diagnosis and clinical decision-making. A successful diagnosis of the movement disorder depends on their ability to transform a material terrain into a meaningful space; a space which both channels and prompts particular interactions within while shielding these interactions from unwanted noise and intrusion. Within such a space, the PT body can perceive and register contrasts to which many other bodies would be insensitive. The resulting knowledge gathered during the GMFM and the musculoskeletal examination help team members to predict how a patient will respond to DBS, thus enabling them to identify which patients are suitable candidates. As I stated earlier, there are no technological methods for doing this. The physiotherapists, however, possess a somatic awareness of how to conduct a successful diagnosis: they have acquired bodies that can register, and be moved by, the subtle differences between various types of motor disturbances. To an important degree the success of the PMDS team depends on these skills. If they cannot screen out those patients who are unlikely to respond positively, and if DBS is too crudely applied to a wide patient group, then DBS in paediatric neurology will, on average, appear to be a clinically ineffective and costly intervention.

Communicative body work and sensorial reflexivity, then, are clearly vital aspects of productive clinical work. Indeed, even in those contexts that are heavily mediated by technology (such as flow cytometry) such skills are necessary to ensure that devices, reagents, samples, texts and so on are correctly arranged and deployed. A small collection of work on the embodied element of diagnostic work illustrates this. Both Goodwin and Moreira, for example, draw attention to the importance of clinicians' kinaesthetic skills in providing important diagnostic knowledge that supplements technology-derived diagnostic information: Goodwin (2010) has illustrated how anaesthetists use their sense of touch to acquire knowledge of aortic aneurisms that is important for the subsequent care of the patient, and Moreira (2006) illustrates how a particular neurosurgical procedure is guided by a neurosurgeon's own touch-derived diagnosis of a patient's blood pressure, combined with the more formally-derived sphygmomanometer measurement provided by the anaesthetist. The case studies provided by these authors illustrate that embodied diagnostic knowledge is an essential element of diagnosis, even in contexts where a great deal of diagnostic work has been delegated to technology. In a similar vein, Schubert (2011) has examined the relationship between diagnostic technologies and embodied skills, illustrating that the latter are configured by diagnostic tools such as the stethoscope. It is by working with tools that clinicians acquire many of the embodied perceptual skills that render them sensitive to otherwise indiscernible elements of the patient. Such tools prompt clinicians to acquire an embodied diagnostic knowledge, without which the tools would be useless and the patient would remain 'unknowable'. Schubert argues that diagnostic tools such as the stethoscope form part of 'diagnostic ensembles in which bodies, tools, and knowledge are mutually configured' (Schubert 2011, 856).

The embodied knowledge of clinicians can therefore be conceptualised as an important element of biomedical platforms. In the case of the PMDS, the embodied knowledge of the PT's is an important part of the proto-platform that has emerged to implement DBS as a therapy for children with dystonia. And as we have seen, such elements form part of a wider array of socio-technical elements that are collectively mobilised, and which configure the productive PT body. Indeed, the GMFM and muscular screen, the material terrain of the hospital gymnasium, such as the guiding floor lines and the adjustable benches and tables, can also be seen as important elements of the PMDS proto-platform. Collectively, they reflect, and perpetuate a particular way of making sense.

Clinician-bodies are an important means by which the clinical gaze becomes embedded. Merleau-Ponty's notion of *corporeal schema*, as articulated by Crossley (2001), is useful here. Crossley describes the corporeal schema as learned embodied 'know-how'. It is a pre-reflective understanding of how to move the body and register the surrounding world: a 'perspectival grasp upon the world from the "point of view" of the body' that is inseparable from practical action and enables reflective thought (Crossley 2001: 102). I suggest that the clinical gaze can be seen in this way. It is perspectival orientation to the world that has become sedimented as part of the corporal schemas of clinicians. It informs, complements, and to some degree responds to a cognitive orientation to the world, such as, for example, a belief that disease should be understood as a biopsychosocial process. What we have seen in this chapter is a specific aspect of the broad clinical gaze of the PMDS. We have focused specifically on the work of the PTs as they instigate a set of practices, during which dystonia is enacted in a specific ways: as an inability to do certain 'gross motor' movements and as particular, bodily movements. This is, of course, one set of dystonia enactments among others that also take place in the PMDS, and these different enactments will involve different forms of communicative body work and sensing-and-acting – it is important to keep this in mind in the following two chapters while we focus on other elements of the platform. As a multidisciplinary team, the PMDS is composed of a heterogeneous group of corporal schemas, some aspects of which will differ according to disciplinary backgrounds, and some of which will be shared (e.g. neurologists and PT's share the embodied knowledge required to conduct a musculoskeletal screen). In this light, we should see the broad clinical gaze within the PMDS as partly being the consequence of this heterogeneous collection of embodied knowledges and skills.

Summary

In this chapter we have examined the way in which the PMDS team manages a specific challenge relating to the clinical adoption of DBS in paediatric dystonia: identifying suitable candidates for DBS. The practices for managing this particular challenge have provided a useful illustration of what I have suggested is an important aspect of the PMDS proto-platform, and indeed biomedical platforms more generally: the embodied knowledge of clinicians. I have suggested that such knowledge can be delineated as involving communicative body work and sensorial reflexivity, and I have suggested that the productive clinical body is configured as such by a carefully constructed clinical space. The PT's carefully construct a material terrain, and then by using communicative body work (in addition to talk) they convert this into a meaningful space that permits particular forms of sensing-and-acting within. Hence, the productive clinician-body is mutually co-configured by other elements of the platform. I have also suggested – in line with the argument of this book – that the clinical gaze becomes both embedded within, and perpetuated by, such embodied knowledge, and hence, within the PMDS, the multiple embodied knowledges of various team members contribute to the consolidation and enactment of a *broad clinical gaze*.

In the next chapter we explore the challenge of managing patient expectations within the PMDS. While doing this we redirect our attention from the embodied knowledge of clinicians and instead focus on another platform element type and thus another way in which the broad clinical gaze is perpetuated. These are elements that align future-orientated visions: when mobilised, these create a set of shared understandings and expectations among clinicians and families about the future. In a general way, this chapter – and indeed Chapter 7 – follows a similar line to this one: we see that the team has adopted a particular tool which serves as a script; we see that, in accordance with the tool, the team carefully arranges a space to encourage particular interactions within; and we see that within these interactions, the broad clinical gaze is enacted and dystonia is rendered intelligible in specific ways.

6

Managing Expectations, Aligning Futures

Families arriving at the PMDS for the first time have varying degrees of knowledge about DBS. Many families have made use of the Internet to access information and arrive at the hospital with 'reams of paper and are very knowledgeable about [DBS],' while other families arrive with 'absolutely no knowledge whatsoever' (Clinical research fellow, interview). Because of the novelty of DBS, however, the information available to patients (and indeed health professionals) concerning the effectiveness of DBS as a means of managing dystonia is limited, and that which does exist relates mainly to adults with the more-responsive primary form of dystonia. Many of those families that do undertake research into DBS have also been influenced by popular news media coverage. Several PMDS children who have undergone DBS, and who have responded extremely well, have featured in regional news media. Again, these are children with the more-responsive primary form of dystonia, and because the media has focused exclusively on those few cases that do extremely well, it can give the impression that all DBS recipients will respond in this way (Racine et al. 2007). This creates a

© The Author(s) 2017   **145**
J. Gardner, *Rethinking the Clinical Gaze*, Health, Technology
and Society, DOI 10.1007/978-3-319-53270-7_6

significant challenge for the team. As the clinical research fellow explains:

> There's often misperceptions, the problem is the press reports the case studies that do well, so there can be perception that DBS will get my child to walk. Because actually that's what you read about in the press . . . We had one child who six weeks after his surgery, was in the press, starting to walk again. And that's a very worrying thing for us . . . Managing the expectations – that can become very difficult . . . there can be [a] perception that DBS will get [their] child to walk

Managing the expectations of families, then, is a considerable challenge for the team. Indeed during my time with the team, it was apparent that a great deal of their discussions about families, and a considerable amount of their interactions with families, were related to this challenge in some way. In this chapter we examine how the PMDS team attempts to manage this challenge and we focus on a novel practice – a goal-setting session – that the team has routinised as a way of re-adjusting families' future-oriented visions. This practice involves mobilising a range of tools, and in particular, a patient-centred tool that has been adopted from occupational therapy: the Canadian Occupational Performance Measure (COPM). The COPM, I illustrate, functions like a script for the goal-setting session. It assigns roles to participating clinicians and families, and in doing so it encodes relations of power. It also represents another way in which the broad clinical gaze is perpetuated within the PMDS: one consequence of its patient-centred focus is that it draws the clinical interest of the PMDS team members to the child's social context and their ability to act within it.

This future-orientated tool is an important element of the PMDS proto-platforms. For the PMDS, such tools and the work they enable are essential for a couple of reasons. First, in an era when facilitating patients' autonomy and enabling patients' capacity for decision-making are heralded as fundamental to ethical medical practice, managing the expectations of patients can be seen as an ethical challenge, and tools such as the COPM enable the team to garner informed consent from patients and families (Gardner and Cribb 2016). In order for families to make an informed decision about whether or not to proceed with the highly invasive procedure, it is necessary for them to comprehend the likely consequences of DBS therapy. The COPM, I illustrate, does this by prompting patients and families to be explicit about their hopes and expectations, and it prompts team members to convey their expectations in terms that are accessible to patients and families.

Second, such tools and the work that they enable help to protect the reputation of the team and indeed of the DBS therapy more generally. It does this by reducing the chances that the PMDS will be accused of failing to deliver an anticipated clinical result. As Dr Martin puts it 'If you paint the most optimistic picture, you are setting yourself up for certain failure'. In the next section I briefly expand on this second point in relation to existing literature on innovation and expectation. We then draw on observation notes and interviews to examine, in detail, a specific goal-setting session. As we do this, we see how future-visions become aligned, and we will see, once again, the broad clinical gaze in action.

Innovation, Expectation, and Promise

In Chapter 2 I suggested that the success of new biomedical technologies and techniques depends on their ability to appeal to, to be co-opted and shaped by, a heterogeneous range of actors. This is of course a future-orientated, promissory process. As a cursory glance of popular news media coverage of biomedical developments illustrates, innovations are the subject of much speculation, calculation, hype, anticipatory coordination, and expectation among potential stakeholders. And as a body work within the 'sociology of expectations' has illustrated, these visions of the future play a performative role: optimistic future-orientated rhetoric animates innovation projects by encouraging the building of alliances (Borup et al. 2006; Brown et al. 2000). By deploying visions of the future and generating promissory expectations, institutions can enrol a potentially diverse array of allies into a common innovation project. Representations of the future have structuring effects on innovation alliances: they delineate and coordinate institutional and professional roles, and by envisaging particular beneficial outcomes and pay-offs, they prescribe responsibilities to those involved. The future of any innovation is, as Brown and colleagues put it, 'a contested object of social and material action' in the present (Brown and Michael 2003, 3).

Much of the social science work in this area has focused on the promissory function of optimistic visions of the future within innovation projects. Obviously such visions (i.e. hype and high expectations) are essential for the attracting the resources, particularly investment, needed to sustain innovation. However, more recently a small body of work has drawn attention to the less-promissory visions of the future that accompany biomedical innovation projects (Tutton 2011; Fitzgerald 2014; Pickersgill 2011; Broer and Pickersgill 2015a; Gardner et al. 2015). This nascent 'sociology of low expectations' scholarship illustrates that the dynamism of some innovation projects emerges from a complex intertwining of low and high expectations; an interplay of optimism, promise and hope, and uncertainty, pessimism and ambivalence. As Fitzgerald has suggested, negative expectations are 'not only thickly present; they may actually be important for the maintenance of some particularly ambiguous neuroscience projects' (Fitzgerald 2014, 2). This aligns with Moreira and Palladino (2005) argument that two logics characterise contemporary biomedicine: a 'regime of hope' and a 'regime of truth'. The former is characterised by the optimistic perception that biomedical innovation is warranted by the promise of a high reward, such as 'health and wealth' generating technological developments. The regime of truth, on the other hand, is characterised by 'the investment in what is positively known, rather than what can be', and the belief that 'most medical therapies are less effective than claimed' (Moreira and Palladino 2005, 67). Biomedical innovation projects are constituted by aggregates and modes of organising that follow either or both of these logics: investors, biotech companies, and prospective patients, for example, are rallied by hope, while regulators and patient support groups may be rallied by 'truth'.

The clinical adoption of promising technologies and techniques represents a particular point of tension between these two logics. One the one hand, hope and optimistic visions are needed to encourage actors to invest time and resources into the adoption and implementation process. Obviously it provides, for example, patients with the impetus to pursue what might be – as in the case of DBS – a highly invasive procedure. On the other hand it is necessary for expectations to be carefully tempered: aside from the ethical implications, constantly missed targets and disappointed patients will discourage further investment of time and resources, and the dissemination of the technology or technique will be in jeopardy. For this reason, tools and practices that help move prospective patients from a 'regime of hope' to a 'regime of truth' are essential components of the adoption process, and indeed biomedical innovation more generally. Hence, the PMDS' goal-setting session that we explore in this chapter has become a vital component of its organisational form, and I suggest we can see the COPM and associated technologies as key elements of the PMDS *proto-platform*. Like other platform elements that we have seen, it participates in configuring clinical work by prompting, constraining, and channelling interaction, and hence, it participates in rendering the patient and their illness intelligible. The tool, as we see here, was developed to reflect patient-centred values: it has been designed to encourage patients to articulate those aspects of their illness that are important to them. Within the PMDS, it compels clinicians to carefully construct personalised, modest, and uncertain visions of the future with their patients. Patients and families are prompted to rationally reflect on their orientation towards the future, and they are encouraged to re-orientate themselves towards an 'expected', mandated future – a future that aligns with a 'regime of truth'.

Informing Families About DBS

Generally by the time families are being considered for DBS they have tried a range of other treatment options without meaningful success. DBS represents something of a last hope for improving the child's quality of life, and their expectations, then, are no doubt heightened due to feelings of vulnerability and desperation (Marks et al. 2009). Caregivers in particular are often highly anxious about the child's future. As one of the PTs stated:

> Parents can come with expectations really high. They want the best for
> their children, and they're also still probably grieving: the emotions

around having a child with long term neuro disability is just unbelievable. It's huge. And they may come in with unrealistic expectations and sometimes DBS is the only thing out there that they could try for that child. And then you're having to wind those down . . . they want their child to talk and [we] are saying we need to rein this down.

'Winding down' expectations, team members said, is a difficult process that requires a lot of time. The long duration of pre-surgical assessments (up to the equivalent of three days) provided such time: team members felt that these assessments enabled a familiarity and a degree of trust to form between them and families, and it provided them with opportunities to continually gauge the capacities of children and their families, and also gauge families' expectations and hopes. There are, in addition to this, two PMDS sessions dedicated specifically to the management of expectations. Here we deal mainly with the second – the goal-setting session – but it is worth saying a few words here about the first: the DBS information session.

The information session is not unlike the standard information session that would be provided to any patient undergoing a highly invasive procedure. It is conducted by the team nurse and includes the patient and supporting family members, and it usually takes place shortly after the patient has been designated as a suitable DBS candidate – although, it is not unusual for some parts of the session to be repeated closer to the time of surgery as there is quite a lot of information to convey. The topics discussed include: details about the DBS device and the surgical procedure, how to take care of surgical wounds, how to recharge the device, how the scars on the scalp and the torso will appear, and how quickly the patient's hair will grow back after the procedure (As the nurse states: 'lots of kids always ask about their hair'). The session is also used to explain and illustrate the efficacy of DBS (as a percentage defined by the literature), and the rates of adverse reactions, including cerebral haemorrhage and infection, and what each adverse reaction would actually entail for the patient: in the case of infection, for example, the DBS system will have to be removed. A PowerPoint presentation with some text and plenty of images (many of previous patients who have undergone DBS) is used to illustrate these various details, and families are given an opportunity to play around with dummy DBS equipment. The session, however, is not used to communicate to families how *their particular child is likely to respond* to DBS. This requires an intense session on its own, after the pre-surgical assessments have been conducted and by which time team members will have a better understanding of the patient's capacity and potential to respond to DBS.

Hence, the goal-setting session takes place after pre-surgical assessments, but ideally a couple of months before the surgical implantation so that patients and families have sufficient time to decide whether to pursue DBS in light of the goals that have been set. The session is essentially a semi-structured interview involving the patient, their supporting family members, and two or three members of the PMDS team. Usually this includes the OT and the PTs who by this stage have spent considerable time with families during the assessments and have been able to form some expectations of their own about how the patient will respond to DBS. In some cases this will also include the psychologist, especially if the patient or family members appear to be particularly anxious. While the main aim of the session is to manage the expectations by setting realistic goals with families, it is also used to quantitatively measure a family's perceptions of their child's abilities, thus providing a means of later assessing whether such goals have been achieved. Usually the session is held in one of the larger consultation rooms, and generally the session takes between an hour and an hour-and-a-half.

The Canadian Occupational Performance Measure (COPM) and Patient-Centredness

The COPM, which provides the basis of the session, is a standardised, semi-structured interview developed in the 1980s by a team of Canadian occupational therapists led by Mary Law. It is designed as a tool for therapists to measure and quantify changes in a patient's functional abilities over time such as before and after a therapeutic intervention. During the interview, patients and their families are asked to identify at least five key tasks of daily living (e.g. washing the dishes and brushing teeth) that the patient would like to improve, and for each task, they are then asked to rate, on a scale from one to ten: the importance of the task to them; their ability to perform that task; and their satisfaction with their ability. From these set of numbers an overall average is calculated. In order to help maintain consistency (and ensure that scores can be compared), users of the COPM are provided with a manual containing a set of instructions and a standardised score sheet (Law et al. 2005).

The COPM, then, measures the patient's self-perception of their abilities. According to the authors, the reason for this focus on self-perception is that the clinically important aspects of an affliction are how it impacts on the actual, day-to-day life of the patient. It is this impact that should be measured (and ideally reduced), and not those aspects deemed important by health professionals who may have very little understanding of the experience of living with the affliction (Law et al. 2005). We can see here how a normative, patient-centred orientation has been incorporated into the tool; an orientation that attaches a heavy weighting to broader 'functionings' of a patient. In this way the tool embodies what the authors describe as the patient-centred mantra that underpins the occupational therapy discipline.

Use of the tool in occupational therapy is widespread, and the authors state that it is now used in over 35 countries (Law et al. 2005). The PMDS, however, appears to be the only DBS service that utilises the tool as part of their routine clinical practice for managing expectations. It was introduced around 18 months after the establishment of the team, largely due to the initiative of the OT. She has used it in her previous work as a community-based OT, and it is, she states: 'embedded in my training'. Its introduction was a novel means of addressing a specific problem:

> [at the beginning] we didn't even know, like these are the first kids to have had DBS, how are we going to set goals?

In the process of being introduced into the PMDS, the COPM has undergone some adaption to suit the aims and local context of the PMDS. First, patients and families are asked to identify the five daily living tasks that are important to them *prior to the goal-setting session*, often very early on during their time with the PMDS. This is so patients can attempt to perform these tasks as part of their pre-surgical assessments (as we will see in the following chapter), and hence, provide team members with an opportunity to gauge how the patient's abilities may improve with DBS. And second, while the COPM in its original form was intended to involve one therapist and one patient, In the team's version, the patient and their supporting family members will be involved (usually both parents), and as mentioned earlier, two or three team members will conduct the session together.

In the following sections, we closely explore a particular session involving Carl (the same patient we followed in the previous chapter) and his mother. The COPM, like all protocols (Berg 1998), provides *a script that coordinates the activities* of the individuals involved. I show that it designates team members as *leaders* of the session, and – in accordance with its patient-centred orientation – it prompts patients and their supporting family members to act as *spokespersons*. Ensuring that patients and family members adhere to the COPM script requires team members to draw on various skills and technologies. In the process, we see another example of the broad clinical gaze of the PMDS, which, in this session, enacts the patient and their affliction as being socially embedded. Towards the end of the chapter, I also reflect on the disciplinary aspect of this gaze, particularly in the way in which it produces a constrained patient voice.

The COPM as a Script

The COPM delegates a 'leadership' role to the therapists, reifying their professional authority during interactions with patients and families. In this role they are responsible for guiding and directing patients and their family members throughout the interaction, prompting them to offer responses that are mandated by the COPM script. This role begins when patients and families are first asked to identify five tasks of daily living that they would like to improve. The COPM manual instructs:

> It is important that clients identify occupations that they want to do in daily life … The therapist should encourage clients to think about a typical day and describe the occupations that they typically do. (Law et al. 2005, 13)

'Leaders' are required to guide patients through a portion of the COPM score sheet that lists a number of possible areas of concern: personal care, functional mobility, community management, household management (cleaning and cooking), play/school, recreations, and socialisation (Law et al. 2005). These 'areas of concern' are used to prompt the family to think about the impact of the patient's motor disorder on their day-to-day life, and for each area, families and patients are asked to identify specific tasks they would like the patient to improve (such as dressing, hygiene, visiting friends, and preparing food). The family is then asked to identify, from these, which five are most important. It is these five tasks that the patient will attempt to perform as part of the pre-surgical assessments (as we see in the following chapter), and that the team will later use as the basis for negotiating expectations during the goal-setting session.

During the actual goal-setting session that takes place after these assessments, the first role of the leaders is to arrange the meeting space so that it is informal and not too intimidating. This is intended to facilitate honest, inclusive, and open communication. In the PMDS, the OT and the PT arrange the space within a large, quiet, and private room so that all those involved in the session are seated in a circle. Here, then, we have another example of a carefully constructed space that is intended to configure particular interactions within.

Second, the therapists begin the session by restating the daily living tasks that had been identified by the patient and their family prior to the pre-surgical assessments. As we see in the following excerpt from Carl's goal-setting session, this is followed by prompting the patient and/or their supporting family members to clarify the specific problem they are having with the task:

OT: Now, when I met you last time, we talked about the things that you wanted to improve. You identified a number of things . . . These were handwriting, shaving, self-feeding, drinking, and using public transport.

PT: So with your handwriting, what aspects are you not happy with? Speed? accuracy?

Carl: Both . . . I get hand cramps. I would like to be able to handwrite on a clear page without making a mark all over the page.
PT: Okay, what about drinking?
OT: I've noticed that when you drink from a bottle, you bring it to your mouth and tip your whole head back. Is that to make sure your arm doesn't flick it away and spill it?
Carl: Yeah.

For each one of the five tasks the patient and their family members are encouraged to be as specific and explicit as possible: the COPM manual states, 'it is essential that therapists use their skills in interviewing, probing for full responses' (Law et al. 2005). As Carl and his mother are prompted to add more and more detail, each particular problem becomes more intelligible and is clearly delineated. In effect, then, the resulting 'five tasks' are the product of a COPM-scripted interaction between the therapists, the patient, and their supporting family members. This interaction may also involve the identification and delineation of a new task. Here is another example from Carl's goal-setting session:

OT: And another thing is you wanted to use public transport. Is this a problem because are you worried that you will attract attention or is it because you cannot physically manage?
Mm: Last time we went out in public, a man laughed at him. The man made such a big deal of pointing Carl out. He hasn't really gone out since. And if he gets one of his tics, it can be very difficult on the bus. They throw him off his feet.
OT: Okay, we should keep the two things seperate: confidence in public and being able to use public transport.

Here we can see that as a result clarification, one of the difficulties identified by Carl and his mother (confidence in public spaces) has been re-delineated as two (confidence in public places and physically being able to use public transport).

Disclosing Predicted Benefits Using Patient-Accessible Frames of Reference

While the five tasks are being clarified, the role of the therapists is to offer their predictions on how the patient's ability to carry out these tasks will be affected by DBS. These predictions are, of course, based on their observations from the pre-surgical assessments and their experience of previous DBS outcomes within the PMDS. If dystonia ('involuntary movement') is deemed to be the cause of the problem, then the therapists will tentatively predict there will be some improvement:

> *PT:*   Carl, tell me about shaving. Why does mum do it for you?
> *Carl:* It pulls on my hair, it is really sore.
> *OT:*   His arm pulls away and the hair gets caught in the shaver. It is definitely the involuntary movements that are making it difficult to shave ... Carl – if DBS does reduce your involuntary movements you will find it easier to shave.

And if a difficulty is perceived by the therapists to be caused by muscle weakness, contractures, or spasticity, then they predict that no improvement will occur. Additionally, the therapists will also draw on their knowledge of previous PMDS patients. Here is the occupational therapist's prediction of how DBS will affect Carl's handwriting:

> *OT:*   I think your computer is your best option. DBS may help a bit, but you won't be able to rely on your handwriting. We have noticed some very minor improvements in patients, but that is after four or five years.

Once the patient and their family have been prompted to clarify the tasks they would like to improve, and once the therapists have offered their prediction, the COPM 'script' requires the therapists to guide the family towards a set of goals for each task. This involves negotiation, and it is during these negotiations that particular goals will be delineated as 'realistic' and others 'unrealistic'. Here is an example of the physiotherapist negotiating with Carl (during which she draws on some of the clinical information that she and the other physiotherapist produced during the GMFM and musculoskeletal screen that we explored in the previous chapter):

PT: About this problem with stability on public transport. We noticed during your [GMFM and muscular-skeletal screen] that you have muscle weakness around your pelvis that DBS won't improve. You could probably improve it with a lot of hard work and exercise in the gym, but we shouldn't set a goal that you are not prepared to put in the effort for in the first place.

Carl: I'm not motivated, but that is because it takes me so much energy to do things!

PT: Can we agree that we don't put this as an initial goal? You could tackle it when you have some more motivation, but I don't think we should put it down as a goal for DBS. We should aim for other goals.

As a result of this negotiation, one potential goal relating to the patient's wish to use public transport has been discarded as an 'unrealistic' expectation.

The following is another example of negotiation, this time involving the psychologist and Carl. (Team members felt that Carl might be depressed, and thus requested that the psychologist participate in the session).

Psyc: Carl, about your wish to have more confidence in public. We need to clarify: What would it take to improve your confidence? Would it be not falling at all? Or falling less?

Carl: Just less falls and less jerky movements.

Psyc: So would just a little bit of improvement, then, help with your confidence do you think?

Carl: Yes.

Psyc: Because some people might not be happy if they still had some visible signs of the movement disorder. It is good that you think that a little improvement will help.

Here the psychologist has implied that it is unrealistic to expect DBS to remove all visible signs of the movement disorder, and has suggested to

Carl that 'a little improvement' is a more suitable goal (i.e. a more realistic expectation).

The product of these types of negotiations, then, is a set of at least five goals, and what exactly counts as 'realistic' or 'unrealistic' are delineated in the process. The resulting five goals, like the 'five tasks' that form the basis of the goals, are a product of the COPM-scripted interactions between participants. If all goes to plan and all participants adhere to the script, the resulting five 'realistic' goals for DBS will reflect the family's wishes for meaningful improvement in the patient's day-to-day functioning, and they will reflect the therapists' educated but tentative predictions: ideally, the five goals will be important to families, and according to the therapists they will be achievable with DBS. Generally, as Carl's aforementioned example illustrates, the resulting realistic goals involve improvements that are modest in nature. The clinical research fellow explains: '[we] find out what their hopes are initially . . . and what we try to do is set some quite modest goals'.

It is in this way that the COPM is used to manage the expectations of patients. The COPM instructs therapists to prompt families to be explicit about their difficulties and hopes, and it prompts the therapists to communicate the effects of DBS in terms that are comprehensible and meaningful to families. Indeed, the potential benefits of DBS are explained in terms of the patient's day-to-day activities: as a potential improved ability to shave with less difficulty, or hold a glass, or prepare a bowl of cereal. According to the clinical research fellow, this is the major advantage of the COPM-scripted goal-setting session:

> This is the idea of doing these goal setting sessions, where you're really critically defining a target, you know, there's no – we're not nebulously talking about things getting better, we're talking about you being able to dress that right arm better. (Clinical research fellow, interview).

A 'well-informed' family, then, is the ideal product of an interaction that has been carefully coordinated according to the COPM. We can see this as a process in which visions of the future are interactionally generated by the various participants. Families are being prompted to rationally reflect on their current limitations and their hopes for the future, and realign their future-oriented visions in accordance with the predictions offered by team members. Thus the futures that are being created are highly specific and tailored to each child and their family, and tend to involve mainly modest improvements. I will come back to this point towards the end of the chapter, but it is important to note that these 'realistic' visions of the future, while highly specific, are also plagued with a sense of uncertainty. Team members emphasis the uncertainty of their predictions, even for modest predictions, when communicating with families. We see this in statements such as: 'DBS *may* help you [with writing], *but*...' And: '*if* DBS does reduce your involuntary movements, you will find it easier to shave'.

With some patients, team members and families are unable to decide on a set of realistic goals. In these cases parents are reluctant to adjust their hopes for the future in line with the visions being enacted in the goal-setting session. When this happens, team members will advise the family not to proceed with DBS. Here the OT provides an example:

[During the session] when you got into more detail, the [father] really wanted a miracle...The dad was like 'I want her to be normal because otherwise I can't give her to get married,' and I said, 'But you know she can do all of the things'. She was really, really functional. She had loads of strength...we didn't put her through DBS because we didn't feel that the goals were realistic and we explained to the family.

Quantification

In addition to managing the expectations of patients the PMDS team also uses the COPM as an outcome measure (as originally intended by its authors). This involves a process of quantification, in which the family's perceptions of the patient's abilities are converted into a set of numbers. If the COPM is carried out again, perhaps after a DBS system has been implanted, these numbers provide a point of comparison and thus a means of measuring the effect of DBS, according to goals that have been established during the goal-setting session.

Once the realistic goals have been negotiated, families are asked by the therapists (guided by the COPM manual) to provide three scores for each of the five daily living tasks which have become the basis of the goals. For each task they must rate on a scale from zero to ten: how important it is to them; how good they think they are at performing the task; and how satisfied they are with this level of performance. The COPM manual recommends that therapists spend time with patients to ensure they understand the rating system, 'using concrete examples such as the judging of figure skating' if necessary (Law et al. 2005, 18). Here is an example of this from Carl's goal-setting session:

> PT: Okay. Let's go on and score these [daily living] tasks ... So firstly, confidence to go out in public places. On a scale from zero to ten, how important is that to you, ten being the most important?
>
> Carl: Eight.
>
> PT: And how good are you at this?
>
> Carl: Zero.
>
> PT: And how satisfied are you with that?
>
> Carl: Two.

Just as the process of goal setting involved negotiation and guidance from the therapists, so too does the process of quantification. Part of the therapist's role as leader is to interrogate patients on the numbers they have chosen, particularly if the rating appears to conflict with earlier discussions. The following is an example of such a negotiation from Carl's goal-setting session. Mum has been asked to provide ratings for Carl's ability to drink:

> OT: Okay, now it's Mum's turn ... Drinking. [Importance]?
>
> Mm: Nine.
>
> OT: Performance?
>
> Mm: Eight.
>
> OT: Really? You hold his cup for him! It is really hard to improve from an eight!
>
> Mm: Oh, okay. Well, I was thinking if him using a bottle. But, I suppose, a two.

Here we see that the quantification of patient's self-perceptions is the result of negotiations between therapists, patients, and family members, in which the therapists act is leaders. Therapists prompt and guide patients and their families towards a set of numbers that are perceived to be appropriate by both parties.

Patients and Family Members as Spokespersons

As Carl's example illustrates, a successful goal-setting session and quantification requires active involvement from patients and families – they are required by the COPM to adopt the role of spokespersons. Spokespersons (or 'spokes-things') are those elements that are designated with the authority to 'speak on behalf of' or delineate and define some entity or state of affairs (Latour 2005, 31). In the process of being instructed, guided, and prompted during the goal-setting session, patients and their family members are called-upon to speak 'with authority' on the impact of the movement disorder on their day-to-day life. The COPM, then, creates space for the voice of families and patients. This is not to say that patients and families can say as they please: they are constantly directed towards the particular modes of expression required for the COPM to function. The COPM script, then, both affirms the importance of responsiveness to patients, and it also configures a particular 'patient voice' and mode of expression.

In PMDS goal-setting sessions, the therapists will attempt to ensure that both the patient and their supporting family members are accorded this spokesperson role. It is not uncommon for a patient and their supporting family members to have different goals and expectations. For younger patients, for example, these revolve around being able to interact with friends, whilst parents are often more concerned about long-term care issues:

> The family might have a lot of care type issues and then the kid will be like, 'Well I don't care about that, I want to access my computer, I want to be able to play with my friends'
> And I think we need to be able to capture that, because the goals are different and that's fine. We might be able to achieve both goals. (Occupational therapist, interview)

Provided both the patient and family members are willing to negotiate a set of 'realistic' goals, such differences are not considered problematic by the therapists. Indeed, in order to engage the patient as much as possible (as stipulated by the COPM) it is common for the therapists to negotiate a set of goals with a patient and then a separate set of goals with supporting family members. In order to encourage patients to answer honestly and ensure they are not 'spoken for' by their family, part of the goal-setting session may be carried out with the patient alone. More often, though, this will simply involve instructing family members to wait their turn and not interrupt one another.

Ensuring that patients and family members adhere to the COPM script and enact this spokesperson role is challenging for team members. It requires them to draw on other tools, their practical skills and their experience-informed dispositions. Often therapists will have to engage in demanding emotional labour (Bolton 2001), during which they must manage their own emotions and those of the patient and family members so that realistic goals can be achieved. This can, for example, entail the use of humour to quell potentially disruptive tension. The following extract from Carl's goal-setting session is an example of this. Towards the end of the session, Carl is visibly tired and his mother is beginning to answer on his behalf. In order to get Carl to provide an answer without being influenced by his mother, she received the following instruction:

OT: Mum! Don't influence him! Hide your face and cover your ears, we need to hear from Carl!

The instruction was given at high volume but with obvious jest. The OT was smiling as she said it, and Carl along with the physiotherapist and psychologist responded with laughter. The instruction had the intended effect: Mum, also chuckling, turned her chair so that she was not facing Carl and covered her ears with her hands, enabling Carl to respond.

Team members also employ the use of communication aids to help patients act as spokespersons. Team members have devised a vision board that can be used to communicate with patients who are unable to verbally communicate or access an electronic aid. Depending on the age of the patient, the vision board will have a series of numbers, the words 'yes' and 'no', or happy and sad faces. As the therapy assistant explains:

> We have a vision board where we put our numbers from one to ten for the importance of a certain goal . . . yes and no smiley face and sad face, in terms of the goals . . . Depends on how able they are . . . we have children who can easily move the upper limb and point.

Such tools, then, provide a means for the therapists to prompt responses from patients with communication difficulties. They function as what Latour refers to as intermediaries (Latour 2005, 39): entities that transport meaning or force from one agent (the patient) to another (the therapists), thus enabling the former to 'act' and influence the latter. In some cases, however, these tools may be insufficient. The severity of the child's affliction may prevent the use of such tools, and their family members will speak on their behalf – although some attempt will be made to include the child:

> If we're unable to have a whole sentence from the child or get an idea what they will want to achieve, then the parents set the goals and we have a yes/no conversation with the child. (Therapy assistant)

These are examples, then, of the type of work that is required to enrol patients and families within the COPM script and configure them as spokespersons. It requires mobilising additional tools such as vision boards and other communication aids, and it requires a tacit social sensibility; an ability to appeal to the emotive capacity of individual patients and family members. We can see this as being part of the necessary labour that is required to interactionally generate 'realistic' visions of the future for patients, and thus bring them within the 'regime of truth'.

Such labour does not always bring about the intended effect. Indeed, in some cases the emotional strain of families can manifest as a breakdown during the session, and inevitably, then, some goal-setting sessions have to be brought to a halt. Here the OT recounts one such case. Their child's dystonia was one symptom of a particularly nasty progressive metabolic condition and the future outlook was grim.

> We were just talking to them, you know, the psychologist and I. Because they knew what was going to happen [to their child] ... [we] made the parents cry. They said 'I don't want to think about it. I know what's going to happen, I just can't get my head around it.

Enacting a Socially Embedded Patient

For the PMDS, the COPM-scripted goal setting session has become a vital part of their routine. It provides an opportunity to encourage patients and families to be explicit about their hopes and expectations, and it prompts team members to communicate their own predictions in a specific manner that is accessible to families. The session, then, is part of their informed consent process, and ideally it helps to protect the reputation of the team and the DBS technique by ensuring that DBS is not oversold to families. The COPM is clearly an important component of the PMDS proto-platform. As with other proto-platform elements, it configures particular forms of patient-centred clinical work, and hence, when mobilised, it participates in enacting the broad clinical gaze of the PMDS.

During the session, patients and their families are encouraged to provide an account of living with the disorder. As we have seen, they are prompted to explain how the disorder impacts upon their day-to-day life in very specific terms, and to identify which aspects of this impact they would most like to improve. They are encouraged to talk about the mundane, everyday life of the child: feeding and drinking, hygiene, sleep, pain, playing with friends, writing and using computers, using public spaces. As a result, during the course of the goal-setting interaction, the movement disorder may be understood as an inability to use the bus, having to rely on mum to help with shaving or drink from a cup, a fear of being laughed at in public by strangers, or an inability to write more than a few sentences before getting cramps. It is according to such details that the movement disorder is rendered intelligible within the goal-setting session. It is, in other words, enacted as social affliction; as a hindrance to engaging and acting in the world of people and things beyond the hospital.

We can see here then how the broad clinical gaze – a gaze that foregrounds, amongst other features, aspects of the patients social context and their ability to act within it – is operationalised via the COPM-mediated session. It is the product of a patient-centred interaction in which team members deploy various tools and skills to ensure that it proceeds 'on script'. Via the guiding and prompting of the therapists, various relations are foregrounded and subjected to clinical scrutiny; relations involving other people (family members, friends, strangers) and objects (computers, pens, cups, buses). The result of this is that the patient is temporally enacted as a social being immersed within a network of relations. This is, of course, future-orientated work. During the interaction, these various relations are foregrounded in order to generate specific but tentative visions of the future for each family, which, as we have seen, tend to involve mostly modest improvements.

It is important to note that the patient's and family member's role as spokespeople is highly configured: they are prompted and channelled towards particular modes of expression as mandated by the COPM. A specific configuration of tools and skills, then, has not only meant that a particular way of engaging (i.e. a particular patient voice) has been produced, it also means that other modes of expression and other ways of engaging and interacting are, in effect, suppressed or elided. What we see here is the enactment of a constrained patient voice (Gardner and Cribb 2016). Here we glimpse the disciplinary power of the broad clinical gaze: it produces individuals who engage and inhabit the world in ways that align with particular socio-technical projects (Rose 2007). This disciplinary dimension is explored more fully in the final chapter, but Oudshoorn's study (Oudshoorn 2015) of patients implanted with cardiac pacemakers and defibrillators is useful here. Oudshoorn illustrates that sustaining these hybrid bodies – or 'cyborgs' – requires patients to possess a certain orientation and sensitivity to themselves and their world. They must be able to sense, for example, a depleted battery so that it can be identified and replaced. Patients are thus active in sustaining themselves as cyborgs, but this only comes about through a disciplining process in which they become attuned to certain phenomena. In a similar fashion, the goal-setting session and the alignment of futures are part of a disciplining process aimed at producing agents with a responsible, rational orientation.

Tools that align visions of the future, and the future-orientated clinical work they enable, are of course features of biomedical platforms more generally. Obvious examples relate to the mobilisation of tools and 'ways of understanding' necessary for clinicians to make a prognosis. As Keating and Cambrosio (2003) note, a driving impetus behind the formation of immunophenotyping platforms was the need to establish correlations between specific biomedical markers and likely disease progression in patients. Biomedical platforms configure forms of clinical work in which a diagnosis is used to project an anticipated set of disease events. Patients are subsequently prompted to align their future-oriented visions and expectations accordingly, and as treatment options and care pathways are planned, patients and clinicians are accorded various roles and responsibilities. Indeed, part of the power of biomedical platforms (and the PMDS proto-platform) derives from their capacity to project and align futures.

In the case of the PMDS, future-orientated tools and the management of expectations is a vital part of the innovation process. I suggest that the goal-setting session can be seen as a mechanism for enrolling families within an innovation network. As I have emphasised throughout this book, the success of a biomedical project depends on the ability of proponents to build and expand socio-technical networks by enrolling other actors. This entails appealing to their interests and goals by deploying optimistic expectations (e.g. hype), and as Latour has argued (Latour 1987, 113–115), by reshuffling and realigning the goals of others, who subsequently feel compelled to join the network. The COPM-scripted goal-setting session does this. It prompts families to move from the 'regime of hope' to the 'regime of truth'; to realign their goals for DBS and thus adopt a future-orientated disposition that aligns with that of team members. Hope and high expectations provide dynamism to biomedical innovation, but by engaging in the sort of work we have seen in this chapter, clinicians are able to discipline this dynamism into specific socio-technical projects, in a manner that is less likely to derail these projects.

Summary

In this chapter we have explored how the team deals with the challenge of managing expectations. Managing the expectations of families is a difficult process that ultimately occurs throughout the whole of their time with the service, but in particular, the team has developed a novel strategy: the COPM-scripted goal-setting session. The COPM has been adopted from occupational therapy (a discipline that prides itself on being patient-centred) and as we saw, patient-centred values have been encoded with the tool. We have seen how the team uses the tool to prompt families to articulate their hopes and expectations, and we have seen how the tool encourages team members to communicate their own, assessment-informed expectations in specific terms. If all goes to plan, team members and families will collectively produce at least five 'realistic' goals, each of which reflects specific tasks of daily living. However, as we also saw, ensuring the families adhere to the script can be difficult. It requires team members to deploy various skills such as the use of humour, and various other technologies such as communication aides. During the session the participants interactionally generate specific, modest, and to some degree uncertain futures for each family. They are relocated with a 'regime of truth', which, I have suggested, is a necessary part of the clinical adoption process.

The goal-setting session is an illustration of the broad clinical gaze at work. We see that during the session, various aspects of the patient's social context are foregrounded and are used to 'make sense' of the illness. I also suggested that the session provided an opportunity to glimpse the disciplinary nature of this gaze. Patients and family members are encouraged to act as spokespersons, but this is a highly configured patient voice. Other ways of engaging and interacting are, in effect, suppressed or elided. In the next chapter, in which we examine how the PMDS measures clinical outcomes, we get another glimpse of this disciplinary aspect of the broad clinical gaze. We again closely examine a tool that has been adopted from occupational therapy and which has been designed to reflect patient-centred values. Like the COPM, the tool also revolves around five activities of daily living that patients identify as being important to them.

7

Measuring Clinical Outcomes

In Chapter 3 we saw how a particular conjunction of historical circumstances provided novel opportunities for managing neurological illness, and we saw how these circumstances brought about some of the infrastructural terrain – the constraining resources – that influence how such opportunities are exploited. The PMDS operates in a climate in which rationalisation has become entrenched via governance mechanisms such as medical device regulation and HTA. Because of this they are under pressure to generate 'objective', quantitative evidence that DBS is a clinically effective and cost-effective technique for managing dystonia in children. For dystonia there is a precedent for how such evidence should be generated: by using the BFM Dystonia Rating Scale clinical assessment tool. This tool was used in the multicentre Medtronic-sponsored trials that were used to obtain regulatory approval for DBS as a treatment of dystonia in both the USA and the European Union, and it has become the standard clinical assessment tool for determining the effectiveness of the therapy, and indeed treatments for dystonia more generally. Because of this, the clinical outcome data reported in the existing body of literature on DBS and dystonia has been generated using the BFM.

© The Author(s) 2017
J. Gardner, *Rethinking the Clinical Gaze*, Health, Technology and Society, DOI 10.1007/978-3-319-53270-7_7

Not surprisingly, then, the BFM is used by the PMDS. If the team wishes to compare their outcomes to those in the literature, or if they wish to demonstrate to external funders and regulatory agencies or the wider neurological community that their therapy is effective, then it is necessary for them to use this standard tool. This, however, is problematic for the team. The BFM tool was developed for use with adult patients with primary dystonia, and it is less useful for patients with the more complex, secondary form of dystonia who represent over half the PMDS patient cohort. Specifically, PMDS team members argue that the tool fails to capture clinical improvements in many of their patients with secondary dystonia; improvements that are relevant and important to patients and families. A concern of the team is that unless such improvements are adequately captured, DBS will, in the future, be withheld from families who could benefit from the technique. As Dr Martin stated in a team meeting:

> If the scale is insufficient then many patients are going to be unfairly excluded from a therapy that can offer them much needed relief. We don't want people to be excluded because an inadequate tool has selectively produced the relevant data, thus giving the impression that only a select few will benefit.

In this chapter I show how the team has attempted to overcome the deficiency of the standard clinical outcome measure by adopting (and adapting) another tool from occupational therapy: the Assessment of Motor and Process Skills (AMPS). Team members argue that the tool is particularly good at capturing and quantifying relevant and meaningful changes in patients with complex movement disorders. By adopting the tool, the team is, in effect, attempting to redefine what counts as evidence within their area of paediatric neurology in such a way that will facilitate further dissemination of DBS.

As I suggest in the following section, the AMPS can thus be seen as key element of the PMDS proto-platform. Clinical assessment tools, I argue, represent the embedding of particular modes of perception, and in the case of the PMDS, the AMPS (along with other tools that we have seen) is another means by which the broad clinical gaze is perpetuated.

In subsequent sections I provide a more detailed account of the short-comings of the BFM as perceived by team members, and then we follow the occupational therapist (OT) as she uses the AMPS tool to measure the effectiveness of DBS. We see that the operationalisation of the tool involves the careful construction of a *domestic space* within the hospital, thus making use of the hospital's patient-centred architectural form. Within this space, we see that the gaze of the clinician is directed towards what could be called the patient's *domestic body technique*.

Conceptualising Clinical Assessment Tools

Measurement is a process of foregrounding and elision. Particular attributes are designated as being in some way representative of an entity of interest, and are thus subject to delineation and quantification, while numerous other attributes are ignored or elided. By doing this, measurement perpetuates a way of making sense of the world according to *differences of degree*. According to this way of making sense, the world is composed of distinct entities that can be compared and contrasted according to the degree to which they possess particular traits (e.g. skin colours, height and length, valency). For theorists such as Deleuze, this 'Platonism' has constraining consequences: it elides the inherent, continuous multiplicities that constitute the world (i.e. *differences in kind*), and perpetuates dualisms that, inevitably, become the basis for normative judgements and corrective measures (Grosz 2005; Hardt 1995). This is of course a key dimension of what Foucault describes as discipline: delineation and measurement, categorisation, normative judgement, and correction. As he illustrates in *Discipline and Punish* (1991), such techniques actively induce and encourage particular ways of inhabiting the world, and indeed it is via the prevalence of such techniques that our 'modern' way of inhabiting the world as individuated, self-responsible agents has emerged. Measurement, then, is not simply a means of describing the world. It is a means of generating and organising a world.

Biomedicine is of course constituted by a plethora of practices that foreground and elide. As Keating and Cambrosio describe it, the emergence of biomedicine, in which biology and medicine have become firmly entwined, has meant that biological parameters are foregrounded as indicative of health or illness, while other patient attributes are passed over. The way in which immunophenotyping platforms are mobilised to measure T-cell populations as a way of assessing the health of individuals with HIV is an example of this. This is the so-called biomolecular gaze of modern medicine (Bell 2013), and as various scholars have argued (Rose 2007; e.g., Rabinow 2008), this is implicated in the emergence of new social forms in which biomedical discourse is incorporated into projects of the self. We must keep in mind, however, that contemporary medicine is far from homogeneous. As the PMDS illustrates, various professional groups, each with their own disciplinary-specific 'ways of thinking' and their own infrastructures may be involved in the frontline of medicine. As we have already seen in regard to physiotherapy (Chapter 5) and occupational therapy (Chapter 6), these groups have their own processes for measuring – for eliding and foregrounding – that are deployed with varying degrees of authority in clinical contexts.

As technologies for measurement, clinical assessment tools should be understood in this way. They tend to reflect the values of the disciplines or medical sub-specialities in which they were developed, and depending on wider contextual factors, they will have varying levels of adoption and implementation within healthcare settings. I suggest that we can see clinical assessment tools as encoding what Goodwin (1994) has referred to as the *professional vision* of the discipline within which they are created. Professional vision is the 'body of practices through which objects of knowledge ... are constructed and shaped' (Goodwin 1994, 605). These include practices of *coding* (delineating objects according to existing classificatory systems) and *highlighting* (or what I have referred to as foregrounding above). Professions, Goodwin argues, are communities that are bounded and animated by their distinct practices of coding and highlighting; practices which provide individuals with a profession-specific conceptual grasp on the world. Clinical assessment tools are, in effect, carefully crafted, standardised practices for coding and highlighting. They do not, then, provide unmediated accounts: rather, they are – to borrow some phrasing from Haraway – *constructed perceptual systems*; highly specific and partial ways of seeing and assessing disease and the body

(Haraway 1988). As such tools are adopted into various clinical and research contexts, these particular modes of perceiving health and illness, these partial, highly specific perceptual systems, become locally embedded. Indeed, as clinical assessment tools become part of clinical infrastructure (i.e. embedded in biomedical platforms), particular understandings of health and illness gain institutional weight.

Within neurological clinical practice, clinical assessment tools can be characterised as existing in one of four groups, according to the way in which they code and foreground/highlight, bodies and body-phenomena. First, there are systems of measurement like the T-cell population counts mentioned earlier, in which biological phenomena (i.e. biomarkers) are used to gauge the underlying disease process. These might include, for example, the measurement of lesions and scarring in the brain as a way of gauging the severity of multiple sclerosis. For many prevalent neurological disorders, however, there are no useful biomarkers for tracing disease severity, particularly during its early stages. Second, therefore, are those assessment tools such as the UPDRS that measure impairment; the degree to which the central nervous system is affected by a disease. Such tools typically involve coding and foregrounding erroneous body movements such as tremors (Martínez-Martín et al. 1994). These two tool types often have their origins in neurology and they tend to reify biomedical model understandings of disease: biological or biochemical elements, or particular clinical signs, are deemed to be 'representative' of 'health'. However, the third and fourth tool types – those that measure disability and those that measure quality of life – have emerged from (or have been heavily influenced by) other health professions that tend to hold less reductive understandings of health and illness. The measurement of disability involves assessing the degree to which a condition reduces or restricts an individual's ability to perform tasks within their physical and social environment. Generally, this is done by measuring the individual's capacity to carryout Activities of Daily Living (ADL), such as eating, washing hands, or self-dressing (Buskens and Van Gijn 2001). The AMPS that we explore in detail further is an example of this, and we can say that it reflects the patient-centred *professional vision* of occupational therapy – the discipline in which it was constructed. QoL-based tools attempt to quantify an individual's overall sense of well-being and life satisfaction.

This is usually done via a questionnaire and/or interviews that prompt patients to report on various physical, social, and emotional aspects of their life – the COPM that we examined in the last chapter is an example of this. The perceived advantage of such tools is that they measure phenomena that patients themselves find meaningful. These tools direct the clinical interest of clinicians to various non-biomedical aspects of health and illness, such as a patient's capacity to act in certain circumstances, or a patient's experiences of illness. In other words, they are an important means by which the clinical gaze is broadened beyond the biomolecular dimensions of health.

Biomedical model-derived assessment tools are ubiquitous and clearly of great value, but wider contextual changes have meant that other measures, particularly QoL-based measures, have also become commonplace (Hobart and Thompson 2001). Armstrong and colleagues (2007) note that emergence of QoL-based assessment tools from the mid-1980s onwards reflected a transformation in medical thinking. It had become apparent in some fields that the biological severity of disease did not necessarily correspond to a patient's experience of that disease nor to their sense of well-being. This realisation, along with the more general rejection of medical paternalism within healthcare, led to the development and dissemination of standardised QoL measures that ultimately derived from health surveys conducted by psychiatrists in the post-war years. Armstrong and colleagues note that the emergence and dissemination of QoL instruments both reflected and crystallised a clinical interest in the distal effects of disease: its impact on day-to-day living (such as impaired mobility or disrupted personal relationships) and its impact on a patient's overall sense of well-being. In effect, Armstrong and colleagues argue, the symptoms of disease were detached from the fleshy body and relocated within the social and psychological domain, and disease, therefore, has been rendered intelligible as having a subjective dimension as an experienced 'illness'. The emergence of QoL-based measures has reified the now common distinction between 'disease' and 'illness'. As these tools have become widespread in clinical contexts, this clinical interest in social and psychological aspects of disease – or this broader clinical gaze – has become consolidated to some degree. This is of course particularly apparent in the management of incurable, chronic disease, in which the aim of clinicians is to manage patients' symptoms in such a way that improves their quality of life.

Governance mechanisms have also consolidated the use of such tools. In the early 1990s, for example, the FDA declared that the efficacy of new neuropharmaceuticals could not be demonstrated with impairment-based tools alone. Clinicians had lobbied the FDA claiming that data derived from impairment-based measures could too easily be 'massaged' by companies to give the impression of clinical efficacy (Buskens and Van Gijn 2001). As a consequence, evidence submitted as part of an application for marketing authorisation must include both impairment-based and QoL-based data. Similarly, HTA bodies in many countries stress the importance of demonstrating *clinical effectiveness* or *clinical benefit*, rather than just clinical efficacy when submitting evidence as part of a cost analysis for new therapies. Clinical benefit will, ideally, capture the real-life improvements experienced by patients, rather than simply their response under highly controlled, clinical trial conditions. Such stipulations encourage the use of clinical assessment tools that capture the broader dimensions of health and well-being.

Various contextual factors, therefore, have prompted a widespread adoption of clinical assessment tools which entail the coding and high-lighting (foregrounding) of various non-biomedical-based aspects of disease. Further on, we see that the PMDS, while retaining the use of an impairment-based measure, has adopted such a tool as part of their proto-platform in order to generate the sort of evidence that is required by existing governance mechanisms. The team's adoption of the AMPS may be unique, but it does represent a wider movement in healthcare, and it is indicative of the increasing influence of the values that underlie patient-centred care. It therefore provides a good opportunity to explore how such tools – *constructed perceptual systems* as I called them – actually participate in enacting disease and the patient. From this we can reflect (as I do in the Chapter 8) on the particular social forms and ways of inhabiting the world that may be encouraged by the broadening of the clinical gaze. As we saw in the last chapter during the COPM, we again glimpse the disciplinary aspect of the broad clinical gaze of the PMDS: during the AMPS assessment, patients are, as with the COPM, provided with a constrained, highly configured space for involvement. This is so that the OTs can establish particular *differences of degree*, specifically relating the patient's domestic body technique.

First, however, the following section explores in more depth the challenge of measuring clinical improvements of children with secondary dystonia, and the perceived shortcomings of the standard, impairment-based measure.

The Burke-Fahn-Marsden (BFM) Dystonia Rating Scale

The BFM Dystonia Rating Scale was developed in the early 1980s by the same group of clinicians that developed the UPDRS (Burke et al. 1985). It contains two sections, the most significant being the 'dystonia movement scale' for quantifying impairment, during which the patient is required to sit upright in a chair and perform a series of basic movements. These include, amongst others: 'Elbows and hands resting on the arms of the chair or the thighs'; 'hold both arms out, extended and supinated ...' 'opening and closing both hands ...' and 'touching the nose with the index finger' (Burke et al. 1985). The patient's performance of each of these tasks is video-recorded by the clinician so that at a later time the performance can be carefully scrutinised and quantified. This involves grading, for each region of the body, which actions exacerbate dystonia (since dystonia can be exacerbated with the initiation of movement), and the severity of that dystonia. By multiplying the two scores together and combining them for each body region, an overall score can be obtained. The BFM also contains a brief disability section during which the patient (or supporting family members) is required to grade their ability to engage in five basic, predefined tasks using a scale from zero to four: speech, feeding, eating, dressing, writing, hygiene, and walking. The BFM will be undertaken with patients as part of a pre-intervention assessment in order to obtain a 'baseline' score, and it will subsequently be repeated at regular intervals after the intervention, enabling the effectiveness of the intervention to be calculated.

Generally, as a scale for primary dystonia, the BFM has received positive reviews within the field of neurology, rating well for validity, reliability, and responsiveness (Burke et al. 1985; Krystkowiak et al. 2007). For the

PMDS, the scale is generally considered to be a useful and effective means of measuring the effects of DBS on patients with primary dystonia. This is not to say that it is without its problems or that it is easy to use. The PMDS therapists who carry out the BFM talk of the considerable work required to get impatient, tired, or playful children to sit and carry out the tasks as instructed,[1] and they refer to what they felt was a lack of specificity in the BFM instructions, which, they state, leads to inconsistent application. While these challenges are relatively minor (and are no doubt common to many clinical assessment tools), there were aspects that could be more problematic, particularly in children with complex movement disorders such as secondary dystonia.

First, the BFM can be difficult to implement because it is not always clear whether movement is directly a result of dystonia alone, or whether it is related to some other component of the patient's condition. Anxiety is common among this group of patients and it can temporarily exacerbate dystonic movements. Patients tend to become more anxious and thus more dystonic when videotaped. Here, the occupational therapist provides an example:

> There was one the other day when the child – it's so funny because he had no dystonia at all during four hours of working with me. But as soon as you put the camera on him, he's like this [mimics involuntary movements]. I say, 'Can you look at the camera?' And he goes [mimics involuntary facial movements]. And I'm like, 'Is that a spasm? That's not a spasm.' But, 'Oh his eyes are closed.' You know, and it ends up being such a meaningless measure for some of the kids.

Second, and more significantly for the team, the tool fails to register improvements that, while they may appear minimal, are nonetheless considered important to patients and their family. This is particularly so with severely affected patients. Here is an example from a team meeting

---

[1] This is what Jesperson and colleagues (2014) have referred to as the care work that is required to produce objective facts; the emotional labour required to ensure the compliance of patients, so that data can be extracted from their bodies.

discussion between the occupational therapist (OT) and a visiting neurologist (NR):

> *OT*:   We have had six cases where there has been no change in their BFM score [after DBS system has been implanted]. But, rather than each patient needing three people to bath them and two to dress them, they now only need one carer. This is a significant improvement, but how do we demonstrate this when it cannot be demonstrated with the available scale?
>
> *NR*:   With one patient, we managed to give him the use of a finger and thumb. This actually translates into quite an improvement in terms of daily living.
>
> *OT*:   That might be so, but how could we demonstrate that to funders?

For a patient, gaining the use of a finger and thumb may have important implications for their day-to-day life; it may, for example, enable them to operate a communication aid and devices such as an iPad. Such improvements, however, cannot be detected with the BFM, and hence it often fails to generate evidence that could be used to justify the use of DBS as a therapy of children with secondary dystonia. With regard to the PMDS patient cohort, the tool therefore fails to meet the 'responsiveness' criteria of 'scientific soundness' requirements of a clinical assessment tool. During an academic meeting, team members discussed whether it was worth dropping the tool from their assessment regime altogether. It was decided, however, that for the meantime they would continue to use the tool as it was nonetheless a useful measure for patients with primary dystonia, and because it had become something of an expected standard within the community:

> *PT*:   So why do we need to use an impairment scale anyway? Is it because all the other research out there uses these scales?
>
> *S&L*:   And the BFM specifically – what is the point of it?
>
> *CF*:   It does enable a common language with other centres.
>
> *Dr M*: The problem is, if we have different scales, different languages for primary dystonia and secondary dystonia, that makes it difficult to compare.

The PMDS, then, has retained the use of the tool, and at this stage it is generally carried out with all capable patients as part of the extensive pre-surgical baseline assessment and at regular intervals after the DBS system has been implanted.

The PMDS is not the only team to note the shortcomings of the BFM. Several other centres working with children with secondary dystonia have also stated that, while it is certainly useful, it can produce potentially misleading data (Marks et al. 2009; Monbaliu et al. 2010; Gimeno et al. 2012). The PMDS' response to this particular challenge, however, is unique. Once again they have drawn on their multidisciplinary expertise and adopted a tool from occupational therapy. The Assessment of Motor and Process Skills (AMPS) is a disability measure; it is intended to assess the patient's ability to conduct ADL. Team members argue that because of this focus on disability rather than impairment, it is more suitable for capturing and measuring improvements that are relevant to patients and their families. They are, in other words, considered more responsive to the effects of the DBS intervention.

The Assessment of Motor and Process Skills (AMPS)

The AMPS was developed in the early 1990s by the occupational therapist Anne Fisher, and it is designed to evaluate the impact of interventions on patients with any form or degree of disability (Fisher 2001). Like the COPM we explored in the previous chapter, it was designed to reflect the patient-centred mantra (or the professional vision) that underpins the occupational therapy discipline: it is intended to evaluate patients' capacity to engage in occupational and domestic activities that patients themselves feel are important (Fisher 2001). The AMPS is generally carried out with patients in the community in their 'natural' domestic setting, but it is nonetheless also designed to be a highly standardised, carefully controlled means of evaluation. Usage of the AMPS is regulated by an organisation called the 'Center for Innovative OT Solutions' that restricts use of the assessment to only those OTs who have become certified by attending a five-day training course. During this time they become calibrated as an assessor, and they receive access to the AMPS software required to conduct the assessment and which can only be 'unlocked' with the OT's calibration code.

A pre-set format for conducting the assessment is provided by an AMPS manual (Fisher 2001). Before the AMPS is conducted the patient must choose several occupational or domestic activities that they feel are important to them (such as brushing teeth, preparing a bowl of cereal, and self-dressing) from a list of 120 standardised 'activities of daily living (ADL)' provided in the AMPS manual. The manual states that these must be tasks that the patient 'wants and needs to perform' but is having difficulty doing so (Fisher 2001). For each ADL the manual provides a description of how the task is to be performed by the patient during the video-recorded assessment, and it describes how that task should subsequently be graded by the OT. The therapist will examine and grade (from 1 to 4) 36 specific skills items that constitute all of the ADL. These include 16 *motor skill items* (such as walking, reaching, and gripping objects) and 20 *process skill items* (such as choosing objects and pacing), all of which are defined by the AMPS manual. Once all items have been scored, the therapist enters the scores into an AMPS software programme, which subsequently computes an overall score for that particular client. During the assessment, then, patients are performing tasks that they feel are important to them, but they are doing so in a manner described by the AMPS manual, and they are being scored according to criteria set out in the manual. The AMPS has become a widespread tool within occupational therapy. Advocates claim that it has been validated for use with patients from various cultural backgrounds (Goto et al. 1996) and with a range of disorders including children with cerebral palsy (Van Zelst et al. 2006). It was while working with children with cerebral palsy in the community that the PMDS OT became familiar with the AMPS, and when she later became part of the newly formed PMDS, she encouraged the team to adopt the tool to help makeup for the deficiencies of the BFM. While the tool has certainly been adopted in other hospital settings (Nygard et al. 1994), the PMDS is the only team using the AMPS to assess the efficacy of a DBS therapy.

A Domestic Assemblage

The assessment, then, requires that the patient choose several daily living tasks from the AMPS manual, and that they then attempt to perform those tasks while being video-recorded so that their performance can be scrutinised and graded. In the PMDS, these are of course the same five tasks that become the basis of the goal-setting session.

In many occupational therapy contexts, the application of the AMPS takes place in the client's home. This provides the occupational therapist with an opportunity to observe a client as they attempt to perform tasks that are important to them, in the environment in which they would usually perform these tasks. As the manual states:

> [OTs] need to observe and evaluate ADL task performances in natural spaces: bedrooms, family or living rooms, kitchens, gardens – ones like those where the client typically would be performing ADL tasks (Fisher and Jones 2010, 4–5)

In the context of the PMDS, where patients and their families may spend several days undergoing an assortment of assessments, it is necessary to recreate these 'natural' spaces within the hospital. As we saw in Chapter 4, the hospital was designed to encourage patient-centred care, and one consequence of this is that it has a variety of spaces, such as the kitchenettes and play areas, that to some extent mimic a normal domestic setting. These provide the OTs with well-equipped settings to undertake AMPS assessments. It is in one of the spaces that we will follow the OT as she conducts the assessment with Carl. Carl is again accompanied by his mother and two of the tasks he chose were washing the dishes and making a sandwich.

Carl's AMPS assessment takes place in a large room with a kitchenette. His mother spends most of the session seated in the corner of the room, knitting, and saying very little. Along one side of the room is a large sliding door that opens to a kitchen sink and dish rack, a fridge, an oven with a stovetop, bench-space and cupboards both under and above the bench. The OT, who is accompanied by the therapy assistant (TA), announces that they will begin with the 'washing the dishes' task. It transpires that the task was chosen because Carl would like to attend university in a few years and this would require him to live away from his mother and clean up after himself.

Just as we saw with the GMFM in Chapter 5, the AMPS takes place within a material terrain that has been carefully arranged by the therapists. This terrain permits the extraction of data from the patient in much the same way as a carefully arranged, controlled laboratory assemblage might permit the extraction of information from an entity of interest (Latour and Woolgar 1979). The assemblage of objects created by the OT is to some extent standardised. The AMPS manual stipulates how the task is to be performed by the patient, and in doing so it designates the types of objects to be included and how they should be arranged within space: indeed, we could say that the description within the AMPS manual is materialised as a specific space, thus (as we will see) enabling the production of useful clinical knowledge. 'Washing the dishes' corresponds to ADL task 'J-2' in the manual, which states that the task must be performed as follows:

> The client is expected to wash and rinse 10 to 15 dishes...Rinsing soap suds off the dishes is expected...Appropriate dishes include an assortment of plates, glasses, silverware, small pans and related utensils. [After the dishes have been cleaned] the client is to drain water from the sink, wipe the counters dry, wring out the dishrag or sponge.

The task description also outlines the responsibilities of the therapist during the performance:

> The client should be completely familiarised with the set-up of the environment including the location of all needed tools and materials (Fisher 2003, 100).

Having checked the manual she has with her, the OT arranges the kitchenette so the task can be carried out correctly. She and the assistant have brought with them a box of dirty dishes from the staff offices on the upper levels. The dishes (bowls, plates, and cutlery) are piled on one side of the sink and the dish rack is placed on the other. The OT checks that there is dishwashing liquid, a sponge, and a tea towel in the cupboard below the sink. The assemblage created by the OT, then, is intended to mimic the usual layout of objects in a normal domestic kitchen – a part of the home is recreated in the hospital.

The OT then familiarises Carl with the ADL task:

> *OT*: Carl – The washing up. How do you normally do it at home? Do you normally do it at home?
>
> *Carl*: No. I don't usually do it.
>
> *OT*: For this one, you are going to rinse and wash the dishes, stack them, dry them, and then put them away, and then you are going to wipe the bench and the sink clean. Got it? Now, you will put the clean plates here and the clean bowls here [in the dish rack] and we will put the cups and utensils over here [next to the dish rack]. You can also put the sponge and the tea towel where they should go when you have finished. Happy?
>
> *Carl*: Yip.

Before he begins, however, the OT instructs Carl to ignore everyone else in the room, particularly his mother. She adds:

> *OT*: Now, I'm not going to interfere with you at all, and neither is mum – isn't that right mum! I will step in, though, if it looks like something is going to fall or something bad like that happens . . . Great. Ready when you are.

Here the OT is attempting to limit which entities can become involved in the domestic assemblage. If it is to yield useful clinical knowledge, then the assemblage needs to be protected from intrusion from unwanted influences that create noise or could throw a carefully coordinated inter-action off course. The built environment of the location where the AMPS is being conducted also reduces unwanted intrusion. In a labora-tory, for example, such intrusion is reduced by the partitioning of space with physical boundaries, thus creating an ordered place of knowledge-production shielded from disorderly outside space (Guggenheim 2012; Henke and Gieryn 2008). The room in which Carl performs the task is already partitioned from the noisy surrounding ward, and with the door closed, the outside world is barely audible. Indeed, those elements and activities that usually constitute a hospital environment are shut out from the AMPS assessment.

Within this ordered space Carl is encouraged to interact with the other elements of the domestic assemblage in much the same way as an 'average' individual would at their home. After receiving his instructions, he stands up and walks to the bench, sighs loudly, and begins to get on with washing the dishes and putting them on the rack one by one. The OT watches closely and takes notes with a pad and pen, and the TA stands nearby recording Carl with a small handheld camcorder. Once all the cleaned dishes have been placed on the rack, Carl rinses the sponge and wipes the bench and sink. Carl appears tired and moves slowly, and his body shakes and jerks due to his movement disorder, but he appears to complete the task without too much trouble.

At the end, the TA has produced a video recording and the OT has created a series of notes. The OT will subsequently use these to closely scrutinise Carl's performance and provide scores for each of the 36 motor and process skill items, thus enabling an overall score to be calculated.

An Assessment of Disability

For each of the motor and process items the OT must assign a number between one and four, where four equates to 'performs skill item readily and consistently', and one equates to 'severe deficit or inability to perform skill item' (Fisher 2003). As an illustration of this scoring process and its implications, we will follow the OT as she uses video recordings to provide scores for a patient who had earlier performed the preparing 'Cold cereal and beverage' ADL task (task C-1). The patient, William, is seven years old and has primary dystonia. William had had the DBS system implanted for about a year, and the assessment was one of several conducted as part of this second post-surgical review.

Before she begins that actual scoring of each skill item, the OT watches the video recording of the performance from start to finish. It begins with the OT giving Williams his instructions: He is to get the milk from the fridge, a glass from the top right cupboard, and use the jug of water that has been provided. The OT shows William where these various objects are. William begins by grabbing a bowl from the cupboard and placing it on the bench. He then picks up the box of cereal and attempts to pour cereal into the bowl. Nothing comes out so William creates a bigger opening in the bag. After several attempts he gets enough cereal in the bowl and uses his fingers to level it. He then gets a large bottle of milk from the fridge and places it in the bench next to the bowl. He stops and looks perplexed, and then pours the milk into the cereal. He over-pours and some milk splashes onto the bench. Using a paper towel he cleans up the mess. Once this is done, he pats the cereal down into the milk with his fingers and then stares at the bowl – he has forgotten what to do next. He asks the OT who instructs him to get a glass of water. He searches the cupboards looking for a glass, locates one, and places it on the bench. He then drags the jug of water across the bench towards himself, lifts it, and then pours the water in. He then stops for several seconds before picking up the glass and pouring some water into the bowl of milk and cereal. He puts the glass down on the bench, pauses, and he then picks it up and pours more water into the into the cereal until the liquid level reaches the very top of the bowl. At this point the recording ends.

The OT then begins scoring one by one each of the 36 motor and process skill items that constitute William's performance. Generally, the scoring process involves an examination of the way in which William uses his body to negotiate and interact with other objects within the domestic assemblage. Each skill item draws the OT's attention to a different aspect of interaction.

For example, several motor and process skills, such as 'Aligns' and 'Positions', relate to the *position of the body* relative to other objects within the domestic assemblage. Body alignment and position are rated according to the degree of functionality: a higher score is achieved if the body is oriented is such a way that the ADL task can be completed efficiently.

If the body alignment results in 'unacceptable delay', 'unacceptable effort', or 'task breakdown', it is scored as a one (Fisher 2003, 189). Here is an example from the scoring of William's performance (as the OT scores each item, she explains her reasoning to me):

> *OT:*   The second [skill item] is 'Aligns' – this is also a four. He had no problem positioning himself to do the task. [Next skill item is] 'Positions' – this is something you would notice when their elbow is way up in the air while they are trying to pour something for instance. There was a bit of that with William. It kind of looks a bit awkward. Lets give him a two.

Some motor and process skill items pertain to specific *interactions between the body and other objects* within the domestic assemblage. For example, 'Grips' refers to the ability of the patient to grasp an object (such as a plate or a pan) or to open containers, 'Lifts' refers to the patient's ability to lift (and not slide) an object from one position to another, and 'Manipulates' refers to dexterity or in-hand manipulation of task objects. Again, body–object interactions are rated according to their functionality:

> *OT:*   'Grips' – there was no grip slip – four. Even if you see a quick grip slip then you give him a two, but there was nothing like that here . . . 'Lifts' – If you slide an object rather than lift it, then you score it down. Did he do it? I think he slid the jug of water across the bench. This is where the video is useful *[the OT consults the recording]*. He did! He gets a two . . . 'Manipulates' – there was some fumbling with the cereal when he was trying to get it out of the box. He gets a two for this.

Body–object interactions are also graded according to their cultural *appropriateness*. The 'Uses' skill item refers to whether or not the correct object has been selected for a task: using "a plate as a plate, knife as a knife'" (Fisher 2003, 207). Here, then, the scoring process reflects social conventions regarding the use of particular objects. This is illustrated in the following extract, where I am asked for my opinion on the appropriateness of an interaction between the body (William's hand) and an object a (bowl of cereal):

> *OT*: 'Uses' – this is about hygiene. For instance, does a stroke victim use a toothbrush to brush their hair? There is an issue with William's hands in the cereal. What do you think?
>
> *JG*: For me, seeing him use his hand to adjust the dry cereal in the bowl looked quite normal. It is something that I would do. But when he used his hand to move the milky cereal, that was different. That seemed unhygienic.
>
> *OT*: Yes, I agree. He gets a two.

Other skill items pertain to the *movement of the body* as the patient navigates other objects within the assemblage. For example:

> *OT*: 'Reaches' – he was fine here. Had no problem getting the milk from the fridge. He gets a four . . . 'Bending' is also a four. There was no increase in effort while picking up the paper towels . . . 'Walks' – no instability when walking to the fridge and back, so he gets a four.

The *fluidity of body movements* while performing the task is also graded. 'Flows' pertains to the 'fluid quality' of arm and hand movement. Smooth movement scores highly, whereas movement that is disrupted, perhaps from 'tremor, stiffness, increased tone' is scored down (Fisher 2003, 196):

> *OT*: 'Flows' – This is hard, I will look [at the manual]. I don't think the spillage was due to 'marked spillage due to tremors', which would qualify as a one in the manual. I think it was due to orientation. So, I will give him a two.

Scoring skill items also involves *differentiating 'erroneous' movements from functional movements*. Movements that are not directed towards the completion of the task or that hinder the completion of the task may result in a lower score. For example, 'wobbly while walking or interacting with task objects' will result on low score of 'Stability' and 'Transports'.

A patient's inappropriate persistence with a task will also be designated as erroneous movement:

> *OT*:   'Terminates' – William gets a one for this. He continued to pour water into the bowl of cereal even when it was full of milk. He did it twice!

Other skill items pertain to the patient's control over their body while moving and interacting with other objects. For example:

> *OT*:   'Calibrates' – Here we are grading his movement and how he adjusts the force of his movement depending on the object. Does he use too much force? Not enough? Well, you can see how much milk he spilled. He gets a one.

The temporal dimension of the performance is also rated. 'Sequences', for example, is scored highly if various steps are performed 'in an effective order for efficient use of time' (Fisher 2003, 214) similarly, 'Gathers' refers to the patient's ability to efficiently collect the necessary items required for a task:

> *OT*:   'Gathers'– He collected everything from the fridge that he needed to in one trip. It would be scored down if he made multiple trips. He gets a four.

And 'Paces' pertains to the overall rate of the task:

> *OT*:   'Paces' – I think he is slow. He is consistent, but he is slow. So, he gets a two.

By the time all 36 skill items have been scored the OT has scrutinised: the positioning of the William's body within the domestic assemblage, the interactions between the body and other elements within the assemblage, body movement and fluidity of movement, and the rate at which sequences of body movements occur. In the next section, I will suggest that this can be considered as a careful examination of what we might call domestic body technique.

Importantly for the PMDS team, the AMPS enables the patient's proficiency in domestic body techniques to be quantified. Once the OT has decided upon scores for all 36 items, she enters them into the AMPS software, along with the ADL task code, the age of the patient, and her personal AMPS code that unlocks the software and ensures the final score will be calibrated according to her severity as a rater. The programme combines these details with scores from several of William's other ADL performances. From these combined scores the programme generates an overall set of scores for William at that point in time. The programme enables the OT to compare these to those of the 'average, able-bodied' seven-year-old child, and with those of William's pre-surgical baseline assessment. The OT notes that William's scores are within the normal range of an average seven-year-old, despite scoring poorly on several skill items. (It is expected, she states, that an average child of that age would have the same difficulties.) And, she adds, when compared with his pre-DBS scores, 'his motor skills have improved significantly, although his process skills are largely unchanged.' It is according to such comparisons of body technique proficiency that that the AMPS enables the impact of the motor disorder, and the effectiveness of DBS as therapy for managing dystonia, to be assessed for each patient.

Domestic Body Technique

The domestic space, carefully constructed in accordance with the AMPS, enables particular, standardised practices of measurement. Specific patient attributes relating to their capacity to engage in culturally mandated domestic activities are foregrounded and scored. Importantly, as William's example illustrates, these various attributes of the performance are rated according to their functionality and efficiency. It is, then, the patient's ability to use their body as an efficient implement that is being examined; an implement for carrying out particular domestic tasks in a culturally permissible manner. To borrow a term from Marcel Mauss (1973[1943]), we could say that the AMPS directs the clinical gaze towards the patient's *domestic body technique*. Mauss defines body techniques as the way in which 'from society to society men know how to use their bodies' (1973, 70). They have, he argues, three characteristics, all of which are scrutinised as part of the AMPS assessment: they involve a specific series of bodily movements and forms; they are social in that they are learned within cultural contexts; and they are efficient in that they serve a specific purpose or function. With clinical assessment tools such as the AMPS that measure disability, health and well-being are enacted as *differences of degree* in domestic body technique. For the PMDS this means that the effectiveness of DBS therapy can be gauged according to quantitative improvements in domestic body technique, relative to the 'average' domestic body technique of someone of equivalent age, and relative to the patient's domestic body technique prior to the intervention.

As with the COPM we see that the operationalisation of the AMPS requires active engagement from the patient. In accordance with the patient-centred values that underlie the tool, patients and their families are encouraged to identify tasks of daily living that are important to them and which they hope DBS will help them to perform better. However this is, as the aforementioned examples illustrate, a highly configured and constrained engagement. The ADL must be chosen from a pre-set list and they must be performed according to a script, as outlined in the AMPS manual. Once again we get a brief glimpse of the constraining and disciplining aspect of patient-centred medicine and its associated broadening of the clinical gaze. As Deleuze and other Bergson-inspired philosophers have argued (Grosz 2005), measurement practices that establish *difference of degree* participate in eliding *differences in kind*: they delineate subjects according to the degree to which they align with certain norms, and hence they mask multiplicity and other forms of heterogeneity. The AMPS requires compliance from patients and it requires certain inputs. It requires a form of purification in which particular features are extracted from complex patient and family experiences of dystonia and DBS; experiences which are no doubt associated with a muddle of strongly felt emotions relating to apprehensions and hopes, uncertainties, and pain. In the following chapter, I reflect on this constraining and disciplining effect of patient-centred practices and the broad clinical gaze in more depth.

As I have also pointed out in this chapter, this constraining and disciplining of patients within practices of measurement is a necessary feature of medical innovation, particularly within the current era of rationalisation. Governance mechanisms such as regulatory bodies and HTA authorities necessitate the production of 'objective' evidence of clinical benefit. By delineating patients according to differences in degree, clinical assessment tools essentially bring them within innovation networks. The elision of patient-complexity and the foregrounding of quantifiable elements enable the patient to be linked to a wider set of relations involving statisticians, commercial institutions, governance agencies, and other actors. Clinical assessment tools have become an essential part of what Clarke and colleagues (2009, 2003) have referred to as the integrated infrastructure of biomedicalisation; infrastructure that enables medical knowledge, technologies, and indeed capital to become ever more co-constituted. As I argued in the exploration of the history of DBS in Chapter 3, clinical assessment tools transform local, contingent, and potential diverse phenomena of particular research sites into a set of easy-to-circulate numbers that can be pooled together, compared, and used as a body of evidence to justify the use of an intervention: they enable the circulation of meaningful data between research sites (the clinic), and between these points and the point at which the efficacy of such interventions are assessed (the governance body). It is also important to note that they extract data from patients in a form that can easily be disseminated and managed with existing information systems and which can be placed in repositories such as patient registries. In short, clinical assessment tools and outcomes measures are one of the crucial technologies that enable populations of individuals to be brought into what Petryna (2006) has called the 'experimental order'.

Hence the AMPS and indeed the BFM represent important elements of the PMDS proto-platform. However, while the widely used BFM provides a common language with other dystonia specialists, the AMPS, due to its relative novelty in neurology, presents an additional challenge. It may capture improvements that patients themselves feel are important, but the team is having to engage in work that demonstrates to other clinicians that the data the AMPS generate do indeed constitute 'evidence' of DBS effectiveness. Some journal editors, for example, may be reluctant to accept manuscripts which present data that has been produced using unfamiliar tools from other disciplines. The PMDS, in other words, is having to redefine what counts as evidence. This can be seen as part of the work that is required to transform a proto-platform into a more widespread platform; a process that is necessary for the further dissemination of the DBS technique in paediatric neurology. We will explore some of this work in the following chapter.

Summary

A major challenge for the PMDS is measuring clinical improvements, particularly in patients with the complex, secondary form of dystonia. While existing heritage tools in neurology such as the BFM have some use for the team, they fail to capture the sometimes subtle clinical improvements that patients and families feel are meaningful. In this chapter, we have explored how the team has attempted to overcome this challenge by adopting the AMPS from occupational therapy. I suggested that like all clinical assessment tools it represents a constructed perceptual system that reflects the professional vision of the discipline within which it was created: in line with the patient-centred mantra that underpins occupational therapy, the tool directs the clinical gaze towards the patient's capacity to perform particular ADL. As we saw, this involves the careful construction of a normal domestic space within the hospital. Dirty dishes, a dish rack, washing liquid, and a cloth are arranged around a kitchenette in an attempt to mimic a domestic kitchen. This space enables an examination of domestic body technique; the patient's ability to use their body in an efficient manner to conduct culturally mediated domestic tasks. The patient – and indeed the clinical effectiveness of DBS – is thus rendered intelligible according to differences of degree in domestic body technique. The resulting data can, then, potentially be used as evidence to support the subsequent dissemination of DBS in paediatric neurology.

Within this chapter I briefly touched upon several themes that are explored in the following chapter, particularly the disciplinary dimension of the broad clinical gaze, and the transformation of proto-platforms to platforms. I address these themes within a more general set of reflections on the social effects of innovation and patient-centred medicine movement.

8

Towards Patient-Centred Platforms

A Brief Recapitulation

At the outset of this book I suggested that if we want to understand the social implications of new technologies in healthcare, it is important to understand how they become embedded in organisational forms. If the potential benefits of a promising technology are to be realised, it will need to become integrated with other technologies, institutional arrangements, and day-to-day workflows. These organisational forms, such as those required to deliver a frontline clinical service, are of course the points at which clinicians, technologies, and patients come together in co-configuring relations. They are the points at which clinicians, technologies, and patients interactionally generate clinical information, shared understandings, expectations and other effects, within which health and illness, patients, and bodies become intelligible. Prognoses are made, futures are aligned, and roles and responsibilities are arranged.

So, in this book I have set about exploring the organisational form that has emerged to deliver DBS in paediatric neurology as a means of

© The Author(s) 2017   **193**
J. Gardner, *Rethinking the Clinical Gaze*, Health, Technology
and Society, DOI 10.1007/978-3-319-53270-7_8

managing primary and secondary dystonia. DBS is particularly interesting to explore because, along with it other neurotechnologies, it is implicated in perpetuating naturalistic, hyper-materialist understandings of health and well-being. Yet, what we see with the DBS is that it has become embedded in an organisational form which has been heavily influenced by values that are commonly associated with patient-centred medicine. Indeed, DBS has become implicated in the perpetuation of what I have defined as the *broad clinical gaze*; a clinical interest that extends from the shapes and structures of the body, to the subjective thoughts and emotional state of the patient, to elements of the patients social context and their ability to act within it. The PMDS as a case study, then, provides an opportunity to explore the development of a novel organisational form around a new technology (DBS), and it provides an opportunity to reflect upon and anticipate the social implications of the patient-centred care movement, and indeed to make some reflections on the impact of new neurotechnologies. Each of these points is addressed in more detail in subsequent sections.

In Chapter 2 I suggested that the adoption of new technologies within organisational forms is a vital part of the innovation process. Common perceptions of innovation as a linear process tend to overlook the creative, pragmatic, and difficult work required of health professionals as new technologies are implemented within clinical services – the *clinical adoption* stage of innovation. A more apt conceptualisation is one that captures the emergent nature of innovation; a conceptualisation in which developments in biomedical understandings, technology transfer, and learning in practice are recognised as important elements of the innovation process. In light of this, I argued that as an analytical frame, we should see the PMDS organisational form as *a proto-platform*; a nascent, socio-technical infrastructure which has emerged to implement DBS in paediatric neurology. The social implications of DBS, then, must be understood in terms of this proto-platform. For this reason, in Chapters 4 to 7 we explored specific elements of this platform, noting how each has emerged as team members have sought to manage particular challenges relating to the implementation of DBS in paediatric neurology. We closely examined platform elements 'in-action' to see how they participated in enacting the broad clinical gaze.

Before doing this, however, we explored the historical development and dissemination of the DBS technique. The trajectory of DBS, as we have seen, has been driven by commercial and professional interests and the flexibility of the DBS technology, and it has been shaped by governance mechanisms, rationalisation, and parallel developments in pharmaceuticals, surgical techniques, and of course, clinical assessment tools. Some elements of the PMDS proto-platform have their heritage in this historical conjunction of circumstances, and we should see the PMDS proto-platform as representing a point at which the DBS innovation trajectory has become entwined with patient-centred care.

The first challenge we explored was coordinating multidisciplinarity; that is, ensuring that the diverse professionals who constitute the PMDS work together as a collective. The PMDS team structure, and the apparently shared understanding among team members about the importance of 'comprehensive' care, can be seen as platform elements. So too can the built environment of the hospital, the payment system, and the carefully crafted PMDS team schedule, all of which participate in facilitating multidisciplinary, 'comprehensive' clinical work. We saw that, as a consequence, team members collectively enacted a broad clinical gaze: during team meeting discussions, patients, and their illnesses are understood according to various biomedical, psychological, and social elements. The second challenge that we explored was identifying suitable candidates for DBS. This requires identifying dystonia and differentiating it from other manifestations of neurological disease such as spasticity so that those patients that are likely to benefit from DBS can be identified. I demonstrated that the team's strategy for overcoming this challenge was to utilise the embodied skills of the physiotherapists. By engaging in communicative body work and utilising carefully honed tactile skills, the physiotherapists are able to identify dystonia and evaluate its impact on the patient's gross motor function. Such embodied knowledge, I argued, is an essential element of platforms. It represents an embedded, perspectival orientation to the world, and for the PMDS proto-platform, it is a means by which the broad clinical gaze is perpetuated.

The third challenge we explored was managing the expectations of patients and family members. Here I demonstrated how the team had created a novel goal-setting session based around an adapted COPM – a tool from the patient-centred influenced discipline of occupational therapy. Provided that those individuals involved in the session adhere to the script, the goal-setting session prompts a particular mode of communication: it prompts patients and family members to be explicit about their hopes, and it encourages team members to convey their expectations in terms that are accessible to patients and families. Hence, the goal-setting session is an important part of the informed consent process, and it helps protect the reputation of the team and the DBS technique. I suggested we see the COPM, and indeed other tools for aligning visions of the future, as important elements of (proto-)platforms. They bring patients and families into a 'regime of truth', and in effect, patients and families become enrolled in innovation-facilitative networks.

In Chapter 7 we examined the challenge of measuring clinical outcomes. Existing heritage tools, as we saw, are of limited use for the PMDS: team members believe that they fail to capture clinical improvements that patients themselves feel are meaningful and important. Because of this, the PMDS has adopted another tool from occupational therapy, the AMPS. Like other platform elements, it directs the gaze of clinicians towards particular non-biomedical features, specifically those relating to the patients domestic body technique. Clinical assessment tools and outcome measures are especially significant elements of platforms. By creating differences in degree, they can produce the 'objectively' verifiable evidence necessary to support further development and dissemination of a therapy. In effect, they bring patients into the 'experimental order'.

We have, then, examined a range of platform elements in our close examination of the PMDS. The resulting proto-platform has been brought about via the creative, pragmatic clinical adoption work of PMDS team members. Such work is transformative: realising the potential of new technologies requires the construction of new elements and linkages to reconfigure existing socio-technical systems. It requires, in other words, the creation of novel assemblages, and in the case of the PMDS these novel assemblages involve a diverse array of tools, bodies, skill sets, fields of knowledge, and elements of the built environment. Indeed, the multidisciplinary nature of the team has enabled it to draw on a diverse array of resources from a range of professions, which together have resulted in a service that is unique within hospital-based medicine and unique within neurology (paediatric or otherwise). The team's use of the COPM and AMPS are good examples of this: both have been adapted somewhat to suit the local clinical context, and both entail clinician–patient interactions that are unusual within tertiary neurological services.

What the resulting PMDS proto-platform elements have in common, whether they be tools for aligning visions of the future, the embodied knowledge of clinicians, ways of thinking, or administrative tools such as the team diary, is that they have structuring effects; they prompt, channel, and constrain the arranging of and interactions between other entities. This is not to say that such elements act in a deterministic fashion. It is by being immersed in a wider configuration of elements – the (proto-)platform – that their structuring capacity is realised, and even then the sensitivity of actors to innumerable other influences will mean that interactions may not proceed on script. Nevertheless, despite this degree of indeterminacy, it is through the embedding of such element configurations as platforms that particular patterns of social forms will emerge. With biomedical platforms such as those described by Keating and Cambrosio (2003) and others (e.g. Bell 2013), the social forms that are emerging are those in which biomolecular features are used as signifiers of health and illness, normal and abnormal. Such platforms are an important means by which the biomedical model of disease is perpetuated, and it has permitted the emergence of social forms in which biomedical discourse becomes the basis for self-understanding (Rose 2007). What we are seeing with the PMDS proto-platform is the initial embedding of particular ways of perceiving and acting upon the patient in which other, non-biomedical features are also used to make sense of health and illness: this is the broad clinical gaze. Obviously, this nascent patient-centred platform is not supplanting existing biomedical platforms. Rather, it sits alongside them, and indeed to some extent they overlap. As Keating and Cambrosio note, new platforms are 'articulated and aligned in complex ways with existing ones' (Keating and Cambrosio 2003, 4).

There are, of course, numerous other PMDS proto-platform elements that we did not explore. We did not explore the cognitive and well-being tests that the psychologist undertakes with patients, nor did we explore the often intense initial consultation with patients in which team members prompt and probe families to remember difficult births and failed developmental milestones. Such histories are used by the team to help diagnose movement disorders. Some of the elements that we could have explored are clearly integrated parts of existing biomedical platforms. We did not explore, for example, the imaging technologies such as MRI and PET that are used to help map the brain of each patient in preparation for the surgical implantation of the DBS electrodes. Neither did we explore the hospital laboratory technologies and protocols that are used to diagnose the metabolic and genetic conditions that underlie some forms of dystonia. Obviously, in these particular settings, the patient, health, and illness are enacted according to biological and biomolecular features. Indeed, as patients (or patient bodily tissue) move between different parts of the hospital, they, and their illness, are temporarily enacted in very particular ways. In those interactions we did explore, we saw dystonia enacted in terms of family well-being, as a hindrance to gross motor function, as frustrated hopes and expectations, and as a hindrance to conducting ADL. There are, then, what Mol (2002) has described as multiple enactments of disease, existing side by side. The significance of the PMDS proto-platform, however, is the way in which these multiple enactments are patched together to enact what we might call a *biopsychosocial* patient. This is what the broad clinical gaze does, it brings together various bio- and pscyho- and social-enactments, each of which is perceived as providing an important, clinically relevant dimension about the patient and their health and illness. Specific practices are involved in patching multiple enactments together. In the PMDS, these include the production of the MDT report for each patient. The report contains the details on medication regimes and the results of the various assessments and other diagnostic procedures, and it contains an overall summary of key clinical considerations, including those relating to biomedical, psychological, and social aspects of the patient's disease. The MDT report will, of course, be consulted and adjusted each time the patient visits the PMDS, and it will also accompany patients as they meet other health professionals and carers in their community setting. It represents an important means by which the biopsychosocial enactment of the patient becomes, to some degree, durable.

In the following section, I will briefly illustrate some of the ways in which the PMDS is attempting to facilitate wider acceptance of DBS as a therapy for movement disorders among the paediatric neurology community. In doing this, we will glimpse some of the innovation work that is required to transform a proto-platform into a more widely spread platform. Then, in the following two sections, I discuss and reflect upon some of our observations of the PMDS and of the broad clinical gaze in more depth. First, I suggest that the patient-centred medicine can be seen as an extension of disciplinary medical power that has both constraining and productive effects. In light of this, I make a normative distinction between totalising and non-totalising disciplinary practices, and suggest that we should encourage patient-centred practices that approximate to the latter. Second, I then use our observations of the PMDS as an opportunity to reflect on the social effects of new neuro-technologies. I suggest that it is important not to overestimate the power of medical technologies and that we should see them as possessing interpretive flexibility. This interpretive flexibility, I suggest, constitutes a space for bioethicists and social scientists to help facilitate responsible implementation and dissemination of technologies such as DBS.

Transforming Proto-Platforms into Platforms

Platforms such as the immunophenotyping platform described by Keating and Cambrosio traverse individual clinical sites. They are constituted by machines, standards, reagents, ways of thinking, and so on that are shared between clinical sites, permitting coordination and communication across geographic space. In Latour's terms, they represent conglomerations of powerful actor-networks; powerful because of their interconnectedness, and because of their relative durability as being institutionally embedded (Latour 1987, 1990). The PMDS, in contrast, is a pioneering service with a relatively novel organisational form, and for this reason I have concep-tualised it as a proto-platform. Drawing on the work of Schot and Geels (2007a) I suggested a proto-platform represents an innovation niche: it represents a carefully crafted space in which a disruptive technology can be initially implemented and refined. The success of the technology depends on the further expansion of this space. It requires, in other words, the dissemination of supporting socio-technical infrastructure, or what I have called proto-platform elements: supporting technologies, ways of thinking, embodied knowledge, and so on. Inevitably, such elements will themselves undergo some modification as they are adopted in local clinical contexts, and in effect, a platform *evolves* from a proto-platform.

The PMDS is involved in various activities promoting the use of DBS within paediatric neurology. It is worth providing an overview of some of these activities here, as they are indicative of the type of innovation work that is required in transforming proto-platforms into platforms. I suggest that we can see this innovation work as creating *durable chains of associations* (Latour 1987; Latour and Woolgar 1979), and also as involving the *acquisition and deployment of reputational esteem*. These are interdependent, but a good example that highlights the former is the team's attempts to promote the acceptance of the AMPS and COPM among the paediatric neurology community. As we saw in the previous chapter, the team believes that a wider array of clinical assessment tools, beyond the well-known BFM, is needed to demonstrate the effectiveness of DBS as a tool for managing complex dystonia. The team, then, must redefine 'what counts as evidence' among a wider neurological community that may be unfamiliar with tools adopted from other disciplines.

The team's method of doing this has been to produce a series of peer-reviewed articles that demonstrate the validity of the tools that they are using. The papers compare and contrast the BFM, AMPS, the COPM, and other tools, drawing on quantitative data derived from the PMDS cohort. In general these articles emphasise the necessity of addressing the concerns of children and their carers, and argue that the BFM on its own is inadequate to do this. In the process, the PMDS is creating a body of peer-reviewed, objective evidence that will be used to justify the use of tools such as the AMPS, and which can be cited when reporting DBS outcomes using non-BFM tools. Members of the wider community might still remain sceptical, but in effect, the PMDS have created an evidence chain that can be mobilised in the face of criticism. They are attempting to create, in other words, what Latour (1987) has referred to as a durable chain of associations, which can provide an authoritative platform to justify their mode of operations and promote DBS.

The team is also hoping to strengthen the evidence chain by participating in the establishment of an international registry for paediatric DBS, known as 'PEDiDBS'. These are registries in which individual patient outcomes are carefully recorded in a standardised format and which can provide an important evidence base, particularly for specialised interventions such as DBS for childhood dystonia, when large-scale clinical trials are not possible (Gliklich et al. 2014). The aim of the PEDiDBS registry, which has taken six years to develop, is to provide an organised platform for data sharing between clinical sites that will ideally lead to 'evidence based practice based guidelines elucidating the role of DBS in paediatric patients' (Marks et al. 2016, 1). The degree to which the registry will specifically support the mode of operation of the PMDS remains to be seen. It is a collective initiative of the small community of clinicians with paediatric DBS experience from around the world: Dr Martin is one member of a small working committee that has been led by clinicians at US-based services, and no doubt the format of the registry (including the type of data that it will capture) will reflect a compromise of concerns of those involved. This is, of course, indicative of how platforms evolve from proto-platforms: elements are adapted and transformed as additional actors are brought in to an innovation network.

The PMDS has, however, positioned itself as an important voice among the small community of clinicians providing paediatric DBS, and indeed among the wider community of paediatric neuromodulation. An illustration of this is the publication of a special issue on paediatric neuromodulation, edited by Dr Martin, in a well-known paediatric neurology journal. The issue contains contributions from leading clinicians from around the world on recent advancements in understandings of neural networks and brain plasticity; various forms of neuromodulation including cochlear implantation, spinal stimulation for chronic pain, and transcutaneous stimulation for bladder and bowel dysfunction; and several papers on deep brain stimulation from French, US, and German-based clinicians. These also include several contributions from the PMDS team, including a paper that compares and contrasts clinical assessment tools for dystonia. The special issue on paediatric neuromodulation is the first of its kind, and Dr Martin's editorial role, along with the strong representation from PMDS team members, is indicative of the PMDS' prominence in the emerging field. Indeed, the PMDS and Dr Martin in particular appear to be held in considerable high esteem by others within the nascent paediatric neuromodulation community. This was apparent at an international symposium on neuromodulation in children that I attended with members of the PMDS. Dr Martin organised the symposium, co-chaired several sessions, and provided concluding remarks, as well as providing a presentation. Several other PMDS team members also presented. During the event it was clear to me that the PMDS had an excellent reputation among an international community of neuromodulation experts. The intention is to have similar symposia at regular intervals every two years.

We can see the special issue and the symposia as illustrations of the PMDS' attempt to create a social network that is likely to be receptive to their mode of operations. Dr Martin in particular has been actively encouraging collaboration among geographically dispersed clinicians and, in doing so, he has no doubt been influential in shaping the direction of discussion. We can see this as ground work that may facilitate the evolution of the PMDS proto-platform into a more widely spread platform. This process involves the *acquisition and deployment of reputational esteem*: Dr Martin, the PMDS, and indeed the UK hospital in which they are based are highly visible and invested with considerable respect, and this no doubt provides additional persuasive power.

It is via such activities (i.e. the construction durable chains of associations, and the acquisition and deployment of reputation esteem) that an innovation niche gradually transforms into a platform. Whether such a paediatric DBS platform will actually emerge is yet to be seen, and if it does emerge (and DBS becomes a routine intervention for childhood dystonia), then it will differ in some significant ways from the proto-platform of the PMDS. Differing budgeting constraints and payment arrangements between countries will mean that willing adopters may have to make do with different team structure arrangements, and this in turn will bear on the types of assessments that can be conducted with patients: not all services would have the luxury of having full-time access to an occupational therapist. As Keating and Cambrosio note, the establishment of a platform is a process of standardisation: it involves collectively negotiating which particular elements need to be conserved across contexts, and which can be regarded as idiosyncratic. This often involves black-boxing essential instruments and interactions within machines such as the flow cytometer, which can then be transported to various clinical sites. Such machines obviously need to be immersed within a wider array of elements to function as intended, but they are nonetheless easier to disseminate than organisational forms such as the PMDS that rely on element configurations that cannot easily be black-boxed. Further on I suggest that this provides an opportunity for bioethicists and social scientists to help shape the emergence of responsible platforms.

Patient-Centred Medicine and the Broad Clinical Gaze

Advocates proclaim that patient-centred clinical services will accommodate the 'whole' patient by being attentive to their unique biography and their subjective experience of illness (Laine and Davidoff 1996). It requires that the clinician 'enter the patient's world, and see the illness through the patient's eyes', so that they can become more responsive to the patient's needs, preferences, and anxieties (McWhinney 1989). It is the operationalisation of the biopsychosocial model of disease (Levenstein et al. 1986), and in contrast to the paternalistic attitude of clinicians in the past, it provides patients with space to contribute to decisions about their treatment and care. Ultimately, advocates claim, it represents a more *humanistic* approach to medicine (Smith 2002).

The PMDS has provided an opportunity to explore patient-centred medicine in practice and thus to critically interrogate these types of claims. We have seen how clinicians can engage with the needs, preferences, and anxieties of patients, we have seen how patients and families can be brought into clinical decision-making process, and from this, we can reflect on what it might mean for medicine to be more 'humanistic'. Just as May and colleagues (May et al. 2006) observed in their exploration of decision-making aids in primary care, we have seen that patient-centred clinical work provides highly configured, constrained spaces for patient involvement. The three interactions that we examined in detail – the examination of gross motor function, the goal-setting session, and the assessment of motor and process skills – are all characterised by moves that, as Jasanoff describes them, involve simplifying the 'messy realities of people's personal attributes and behaviours and converting them into [an] objective, tractable [discourse]' (Jasanoff 2004, 28). There is a form of purification occurring in which particular bits of information, related to pre-set bio, psycho, and social criteria, are extracted from messy, cumbersome personal detail. This entails the production of both quantitative and qualitative data, and consequently the establishment of *differences of degree*, in which patients can be compared before and after DBS implantation, compared to one another, and compared to the 'average child' of the equivalent age. This, I suggested, reflects a disciplinary aspect of the broad clinical gaze. During clinical interactions, patients and families are prompted to follow pre-set scripts; to proffer a particular patient voice, and to align their visions of the future with those rational and realistic visions of the team. Patients and families cannot say and do as they please: they are, in other words, disciplined into an innovation assemblage.

The broad clinical gaze, then, is normative. During clinical interactions, patients and families are encouraged to approximate a set of ideals about how patients should engage with their care. These ideals reflect the valorisation of a particular type of citizenship based on the 'empowered' individual – or in the case of the PMDS – 'empowered' families: informed, health-seeking, and responsible families with a rational orientation to their future. Here is, as others (e.g. Thompson 2007) have noted, an alignment with the ethic of responsibility and consumerism that characterises late modernity. What we see in this example of patient-centred care, in other words, is an example of biopolitics (Rose 2001) in which healthcare practices become the means for prompting a particular form of self- and family governance. Perhaps *biopsychosocial-politics* is a more apt term to describe this mode of governance. As we saw during team meeting discussions outlined in Chapter 4, patients are inevitably scrutinised according to a broad variety of norms relating to psycho, social, and biological attributes. Team members would make judgements about age-appropriate behaviour, family dynamics, and relationships with peers, and so on, and clinical decisions were sometimes orientated to uphold such norms. In this light, a 'humanistic' approach to medicine – one that reflects the biopsychosocial model – entails what could be described as an extension of medical power, from the purely biomedical realm to aspects of the psychological and social realms. This is perhaps most stark in the assessment of motor and process skills (AMPS), in which we see medical power extended into the domestic life of patients and their families.

However, we need to be cautious not to overestimate the power of the specific broad clinical gaze of the PMDS. Ultimately, as Rose notes, disciplinary practices acquire their potency as individuals themselves actively and creatively incorporate them into techniques of the self. Patients and families spend considerable time with the PMDS during pre-surgical and post-surgical assessments (which are repeated at regular intervals), but we can only speculate as to whether their participation in these patient-centred practices actually encourages forms of self-awareness and practices of governance in their day-to-day life. It may be that the experience of being carefully scrutinised according to a broad array of norms is quickly forgotten by families as they return to the messy toil of the everyday – although in likelihood, depending on how the MDT report is used by community health professionals, they will be subject to some ongoing scrutiny. Nevertheless, as an example of a patient-centred service, the PMDS does provide a useful point of reference for anticipating the implications of the patient-centred movement more generally.

This is particularly important given that the patient-centred care is widely advocated in healthcare policy in many countries as a way of improving patient outcomes and making healthcare services more efficient (e.g. Epstein and Street 2011a; DoH 2003). Currently in countries such as the UK, there is a marked gap between policy discourse proclaiming patient-centred principles and healthcare as it is practiced 'on the ground'. The way in which such principles are and will be operationalised will vary across contexts, depending on local infrastructure and expertise and the nature of the disease (chronic or acute) being managed (Cribb 2011). But, 'patient-centredness' can be seen as a potentially powerful set of ideals that are functioning as an organising principle for inducing change. An example of this is the NHS England's 'House of Care' model. Launched in 2013, the House of Care model is described as an initiative to facilitate the changes necessary for coordinated patient-centred care for those with chronic conditions, such as cardiovascular disease, diabetes, and other disabilities. The 'House' represents the four key interdependent components needed to achieve this. As its base, it has a commissioning and budgeting system that supports two 'walls': engaged and informed individuals and carers; and healthcare professionals committed to working in partnerships. These, in turn, will entail changes to organisational arrangements (the roof) so that they are 'structured around the needs of patients' (NHSE 2016). Hence, the intention of the initiative, according to NHS England, is to prompt the system-wide change that is needed to bring about patient-centred care. Ensuring that patients are 'truly empowered … requires a massive cultural shift' (McShane 2013), and the House of Care model will help coordinate the action so that such a shift can occur. We can see it is an organising principle or a policy technology intended to create patient-centred socio-technical infrastructures within the NHS, perhaps similar in some respects to those of the PMDS. Indeed, it ties in with another initiative with a similar aim, the Integrated Care and Support project. This is intended to ensure a seamless service 'focused on the individual' that overcomes a perceived divide between health and social care. A major motivation for the Integrated Care and Support project is to reduce overall hospital admissions by promoting greater coordination between the hospital, GPs, allied health professionals, and carers, so that a seamless service will extend into 'their own home' (DoH 2013, 1). Similar to the PMDS, this entails a degree of multidisciplinary coordination around the patient, and the permeation of the clinical scrutiny into their domestic sphere.

Just what effect NHS England's House of Care model and the Integrated Care and Support initiative will have on existing clinical and social services remains to be seen, but we can see them as a possible means by which a broad clinical gaze becomes further institutionalised within healthcare systems. If such initiatives are successful in bringing about the intended changes, then a significant number of individuals and families will find themselves immersed in practices in which individuals are enacted as biopsychosocial agents with mutable (i.e. correctable) bio, psycho, and social characteristics. Within such practices, a broader range of experts would be enrolled into a biopolitical project (Gardner 2016b): Various allied health professionals – physiotherapists, occupational therapists, speech and language therapists, psychologists – in addition to medical doctors, would be endowed with the authority to speak about, delineate, and attempt to correct some aspect of the individual, while patient engagement, as we have seen in the PMDS, could be highly constrained. We might reasonably imagine that as such patient-centred practices become increasingly prevalent, individuals themselves would be more likely to actively incorporate the discourse and corrective measures of these professionals into practices of the self, pragmatically and creatively combining them with other belief systems and ways of understanding.

Patient-centred initiatives are therefore an extension of medical power. It is a power that is constituted by a wider range of experts, and as we have seen, by various socio-technical elements such as patient-centred tools, architecture, and payment systems. This medical power, however, need not be seen as a dominating force that subjugates passive patients and families. The broad clinical gaze has disciplinary, normative effects, but as Foucault (1991) makes clear, while disciplining can prohibit and elide ways of existing in the world, it can also create affordances for the disciplined individual. A disciplined individual is one who has become carefully attuned to certain aspects of the world and their inhabitation of the world. They have acquired, as Latour (2004) puts it, a capacity to be attuned to, and affected by, elements that they would otherwise be insensitive to. The physiotherapists that we followed in Chapter 5 are a useful example of this: disciplinary techniques (professional training) have produced a corporeal sensitivity that functions as a type of capital for the physiotherapists. It permits participation in networks (such as professional groupings) and it provides a means of self-understanding and identity. Indeed, this is what it means to say that disciplinary power has productive effects: disciplining precludes potential modes of existing in the world, but it opens up other trajectories of experiences which individuals can pragmatically incorporate into practices of the self. If we are to make normative judgements about patient-centred practices and the broad clinical gaze, then we need to keep in mind this dual constraining/productive capacity of discipline.

At this point, I suggest that we can make a normative distinction between totalising disciplinary practices and those that are not (or a totalising gaze, and those that are not). This distinction between the two – which can be seen as representing two ends of a continuum – provides an additional means of making a judgement about the moral value of patient-centred medical practices such as those that constitute the broad clinical gaze. Totalising disciplinary practices represent a scenario akin to a total institution. Individuals are immersed in rigid assemblages that permit certain foregrounding/elision practices only. Individuals become intelligible according to specific differences in degree, and the range of experiences open to them, and the opportunities for self-affirmation and self-understanding, are severally limited. A real-life clinical example that comes close to this is Latimer's (1997) account of an acute medical unit in which elderly individuals were configured as particular types of patients with limited futures, and thus to be 'disposed of' by organisational routines that provided no space for patient input. In such circumstances, the individual is dominated by the practices within which they are immersed. In contrast, when individuals can move between different disciplinary practices, each providing different experiences and means of self-understanding, we can describe such practices as having liberatory consequences. In making this distinction and valuing the two situations in this way, I am aligning with a Deleuzian-inspired moral framework articulated by Nick Fox (2012; 2013). Fox argues that well-being and health should be understood as a capacity of a body to be immersed in and move between various assemblages. Assemblages (which are constituted by other bodies, tools, and ideas) affect and discipline the body in different ways, providing new experiences and permitting forms of action and creativity – obvious examples include the assemblages entailed in music lessons, yoga, carpentry, or surgical training. Fox suggests that ill health should be understood as a reduced capacity to engage in such assemblages; it limits the capacity to experience, to act, and to be creative (Fox 2013). Health and social care, therefore, should aim to facilitate the body's capacity to engage. This not only entails clinical interventions to enable mobility, for example; it may also mean other types of therapies, such as art and music therapy, which can provide the individual with new resources for reflection, affirmation, and creative engagement.

In light of this moral distinction between totalising and non-totalising disciplinary practices, I end this section with two normative reflections. The first of these is that patient-centred medicine and the broad clinical gaze can have liberatory effects. The PMDS provides a useful example of this. Due to the complexity and severity of their movement disorders, many PMDS patients have a limited capacity to move between assemblages, and their ability to engage with and experience the world is severally restricted. The broad clinical gaze, by encouraging an intense awareness of particular bio, psycho, and social aspects, may open up new trajectories of experiences which enrich the therapeutic effects of DBS. We might imagine, for example, that after conducting the AMPS some patients become intensely aware of their domestic body technique, and if DBS provides them with greater mobility, they may then consciously aim to improve their AMPS score and refine their domestic body technique within 'practices of the self'. This would be an example of the productive power of the broad clinical gaze. However – and this is the second reflection – the liberatory effects of the broad clinical gaze will be lost if it becomes near-totalising. Patient-centred practices, as we have seen, provide a highly constrained voice for patients and their families, and the authority to speak of the disease is delegated to various health professionals. If particular patient-centred infrastructures become widespread, and if patients find themselves locked in such practices, then their capacity to experience themselves and their world will be restricted. The biopsychosocial model that underlies patient-centred medicine could become something of a straightjacket. As patient-centred medicine is increasingly championed as something of a panacea to current challenges in the healthcare system, the risk here is that 'good and comprehensive' care will *only* be equated with those practices that 'empower' families to be informed, to be responsible, and to obtain a rational orientation to the future. There needs to be protected spaces for other care practices which permit and encourage other forms of engagement in the world. These may include art and music therapies, but ideally, care practices will also entail something akin to what Latimer (2014) refers to as *careful science*. This, as Latimer describes it, occurs when health professionals have the capacity and willingness to be affected by other-ness, heterogeneity, and uncertainty[1] while working with patients and families. We might also describe careful science as interactions in which patients and families are rendered intelligible in ways other than accord-ing to pre-set *differences of degree*. This is what is required if health professionals are to do as patient-centred medicine advocates claim and 'enter the patient's world … see the illness through the patient's eyes' and thus become more responsive to the patient's needs, prefer-ences, and anxieties (McWhinney 1989).

Neurotechnologies, Neurosociality, and the Rise of Neurocentrism?

We see clearly with the PMDS that the development and dissemination of deep brain stimulation *does not necessarily entail the reification of brain-based explanations of the self.* In some aspects, this observation is at odds with other sociological studies on new medical technologies.

Keating and Cambrosio's account of the emergence of biomedical plat-forms aligns with much of the work in the social sciences exploring new medical technologies (e.g. Brown and Webster 2004; Clarke et al. 2009; Corea 1985; Klein et al. 2013). As Webster (2002) summarises, many new technologies enable the medical gaze to peer deeper into the body, thus facilitating the emergence of new health identities based on biomedical criteria. Rabinow's (2008) term *biosociality* is often used to capture these social forms in which biomedical discourse provides a means for organising social life (Rose and Novas 2004; Novas 2006). As Rose puts it, the power of contemporary biomedicine means that our sense of individuality increas-ingly derives from our knowledge of biomedical traits, and is then 'increas-ingly grounded within our fleshy corporal existence' (Rose 2007, 26). Scholars have also examined what we might call *neurosociality*: a form of biosociality in which neuroscience-derived understandings of brain

---

[1] Latimer's notion of 'careful science' will be explored in more depth in her forthcoming book: *Aging and Biopolitics at the Limits of Life*.

networks, brain chemistry, and brain functioning are drawn upon by actors to make sense of health and illness, and are increasingly used by actors as a basis for self-understanding.

As Singh (2013) states, new neurotherapies and neuro-imaging technologies have been implicated in the emergence of brain-based explanations of health, illness, and personhood – or what some authors have referred to as the creation of *cerebral subjects* (Ortega 2009; Vidal 2009). Indeed, a significant body of social science work has set about exploring the recent rise of neuroscience and its cultural impact (e.g. Pickersgill and Van Keulen 2011; Rose and Abi-Rached 2013; Joyce 2008; Dumit 2004). As Pickersgill and Van Keulen (2011) argue, the neurosciences have emerged as an influential and prestigious branch of biomedicine, and we can see this illustrated in the considerable sums of money that have been directed to vast multi-institutional projects, such as the US National Institutes of Health's BRAIN Initiative, and the EU's Human Brain Project. Neuroscientific models and neuroscience tools (particularly neuro-imaging technologies) are increasingly influencing the way in which the 'classical questions' – such as the nature of human agency and the demarcation between normal and pathological – are being addressed (Pickersgill and Van Keulen 2011, xii). Individual behaviour, identity, and the self, in other words, are increasingly being rendered intelligible according to the brain and its structures.

Sociological inquiries have thus sought to examine different manifestations of this neurosociality. Rose (2003), for example, has traced the emergence of the neurochemical self in which individuals are rendered intelligible in terms of their brain chemistry. According to this understanding, the brain – and thus the individual – is malleable and open to improvement via carefully administered psychopharmacology. Indeed, this understanding of the individual is implied and perpetuated by psychopharmacological interventions. Fein (2011) has highlighted a potentially contrasting neurosocial enactment of the self: the neurostructural self. Drawing on her study of people diagnosed with an autism spectrum disorder, she argues that this particular means of understanding personhood equates the self with a brain that is largely a fixed material system governed by physical laws. The neurostructural self is predictable, and in contrast to the neurochemical self, it is 'not open to intervention and optimisation' (Fein 2011, 27). Along similar lines, Rapp (2012) has noted how families of children diagnosed with ADHD and learning disorders draw on neuroscience- informed, brain-based explanations of the self to make sense of their child's behaviour. Ortega (2009) has also noted a similar appropriation of neuroscience discourse among social groups involved in autism advocacy. These groups made brain-based distinctions between 'neurotypicals' and those with autism, and argued that both were examples of natural neurodiversity, thus problematizing common perceptions of autism as a pathology.

There is, then, a diversity of emerging neurosocial forms, but generally they reflect a prevalent shift towards biomedical, brain-based explanations of disease and the self. As Rapp argues, 'psychodynamic explanations of human variation and suffering' are being eclipsed by 'brain-orientated, hyper-materialist explanations' (Rapp 2012, 8). DBS is implicated in this 'march towards neurocentrism'. This is of course to be expected, given that it targets the structures of the brain to treat various forms of psychiatric or motor abnormalities. Moutaud's (2012; 2016) ethnographic study of an adult DBS service – which I briefly reviewed in the introduction – illustrates this. He observed that patients with OCD undergoing DBS understood their illness as a manifestation of problematic neuro-circuitry, and along with Parkinson's patients, they tended to attribute some aspects of their behaviour to DBS and its direct effects on the brain. Within these contexts, Moutaud suggests Parkinson's and OCD patients are to some extent buying into materialistic, brain-based explanations of behaviour and the self. Hence, given that the PMDS puts electrodes into the brains of their patients, and given that they utilise neuroimaging technologies such as MRI and PET (known to reify neurocentric understandings (Joyce 2008; Dumit 2004)), we might expect the PMDS to also deploy 'brain-orientated, hyper-materialist explanations' of their patients. Within the organisational form of the PMDS, brain structure and function are obviously vital dimensions of health and illness, but as we have seen, DBS has become implicated in the perpetuation of a broader, biopsychosocial understanding of health, illness, and personhood. For the PMDS, DBS is understood as a tool for 'unlocking' the body and enabling patients to engage in the world as biological, psychological, and social beings. Depending on if and how the PMDS proto-platform evolves into a more widely spread platform, this biopsychosocial manifestation of neurosociality may be propagated in other contexts as DBS is more widely implemented.

We should be cautious, then, about attributing particular social effects to new technologies without examining the organisational forms within which they become embedded. They possess, of course, what STS theorists have referred to as *interpretive flexibility* (Bijker et al. 1987). As technologies disseminate, they are adapted to align with local aims and interests, and they become immersed within local socio-technical infrastructures that entail particular ways of making sense of health, illness, and the patient. It is useful here to draw a parallel between the interpretive flexibility of neurotechnologies, and the interpretive flexibility of neuroscience discourse as examined by Pickersgill, Broer, and colleagues. These authors have carefully interrogated the diffusion of neuroscience discourse among the public (Pickersgill 2009; Pickersgill et al. 2011) and among policy makers (Broer and Pickersgill 2015a, b). In their focus group research with members of the public, for example, they note that neuroscience discourse is pragmatically adopted in varying and sometimes creative ways. When prompted by the researchers to account for behaviour and action, some individuals appeared to express brain-based explanations, and others drew on more sociological explanations, attributing an individual's behaviour to their upbringing and environmental context. Indeed, some participants deployed a neurodevelopmental narrative that aligned with that of PMDS: Personhood was not simply equated with the brain: it was seen as a product of a person's childhood social ecology – the 'multidimensional web of actors, structures and experiences within which they [were] situated' (Pickersgill 2009, 58). Pickersgill and colleagues suggests, then, that people make sense of themselves and the world around them by 'diverse knowledges pertaining to soma, psyche and society' (Pickersgill et al. 2011, 346). Hence, in an era where neuroscience has considerable authority and esteem, and when neuroscience discourse is being widely circulated, we find that multiple neurosocial understandings are emerging among different groups of people. We can also think about the dissemination and adoption of new neurotechnologies such as DBS in this way. DBS has been pragmatically incorporated into local systems of understanding in several clinical contexts, and it is therefore facilitating multiple understandings of health and illness. These may often be, as Moutaud's study illustrates, reductive, brain-based understandings. We might also imagine that the utilisation of DBS for psychiatric disorders will involve the enactment of highly reductive, biomedical understandings of conditions such as depression and anorexia; conditions that might otherwise be understood according to personal histories and social contexts (although how exactly these conditions are rendered intelligible will depend on the other interconnected elements of the emerging proto-platforms)

Recognising the interpretive flexibility of technologies cautions us from over-estimating their power. Their transformative social effects depend not on their intrinsic qualities, but no on how they are worked into specific practices by creative, pragmatic agents. This has important implications for bioethicists and social scientists. It means that we can encourage the formation of platforms that make use of promising neurointerventions while also perpetuating understandings of health and illness that reflect patient-centred values – if it is deemed desirable to do so. We might, for example, encourage the provision of resources that would permit adult DBS services to adopt some elements of the PMDS proto-platform, such as the AMPS or the COPM. We can discourage organisational forms that may entail disciplinary practices that are near-totalising, and encourage those that facilitate the multiple trajectories of experience – although identifying such forms would require close empirical study. The flexibility of technologies such as DBS, in other words, provides us with a space to promote the use of supporting tools that would help facilitate good ethical work. As I have suggested elsewhere (e.g. Gardner 2016a; Gardner and Cribb 2016) this would constitute a form of ethical work in which bioethicists or social scientists would become more like architects or engineers, seeking optimum platform arrangements – in conjunction with biomedical workers and patients – of tools and objects to help 'construct' a favourable ethical terrain. In effect, then, we would be participating in shaping the evolution of proto-platforms into more widely spread platforms.

Notes

2. Understanding Innovation and the Problem of Technology Adoption

A methodology that aims to explore emerging medical platforms and their social implications requires an approach in which the researcher can witness day-to-day clinical work first hand. Ideally, this would involve prolonged immersion in the field site using ethnographic methods, but in contemporary healthcare contexts, access is often severally limited. This research was undertaken in the UK's NHS, and this required that it be approved by a relevant NHS Research Ethics Committee. A stipulation of the approval was that it would be necessary to require informed consent from all participants (and assent from all participants under 16) involved in an observed interaction prior to that interaction actually taking place – it was not permissible, then, to simply shadow clinicians and collect data as they went about their day-to-day work. For this reason, I developed the following strategy: I spent the first few months of the fieldwork observing regular team meetings which I used to help identify specific challenges that were being grappled with by the team. I then requested permission (from the clinicians, patients, and

© The Author(s) 2017   **215**
J. Gardner, *Rethinking the Clinical Gaze*, Health, Technology
and Society, DOI 10.1007/978-3-319-53270-7

accompanying family members) to observe clinical interactions in which such challenges were being encountered and managed. While doing this, and while continuing to attend regular team meetings, I also interviewed each member of the team (some of them twice), during which I was able to ask them to reflect upon things that I had witnessed during my observations. Team members also took time to demonstrate other relevant aspects of their work (such as how they prepare the team schedule, which is explored in Chapter 4), and I attended several colloquiums in which team members were heavily involved. In total, I observed around 25 team meetings, I conducted 12 interviews, and I observed 7 clinical interactions.

All interviews were audio-recorded, but it was not possible to make audio-recordings of observations. For these, I followed the written note-taking strategy recommended by Hammersley and Atkinson (2007). In order to record as much activity as possible, in as much concrete detail as possible, I created my field notes in two stages. The first stage took place during the observation. Here, I made very concise notes comprising key words and phrases and brief descriptions: 'jottings, snatched in the course of action' (Hammersley and Atkinson 2007, 143). These were then used to trigger my memory during the second stage of note taking that took place immediately after the observation when I wrote up a more substantial and detailed reconstruction of the observed interaction (stage two). However, in addition to capturing what was 'said' by participants, it was also necessary to record as much of the non-verbal action as possible, and thus capture the interactions between bodies, objects, technologies, and the physical built-environment. Following the design theorist John Zeisel (2006), I created annotated maps of the physical setting in which the observed scenes take place. During the interaction I created a quick floor plan of the setting, noting the location of tables and chairs, the position of various participants (PMDS team members, patients, family members), and the position of various equipment and tools that were used. I was particularly attentive to what Zeisel refers to as *adaptations for use*: the way in which participants adapt their material context so that it can be used in a specific way. The reader will note that Chapter 4, and particularly Chapters 5 to 7 are heavily indebted to this method of data collection.

The analysis of data was an iterative process and took place throughout data collection. All interview transcripts and field notes were transferred to NVivo9 data storage/analysis software, which was used for data coding. After five of six observations of team meetings I had identified what I took to be the major challenges of the team: coordinating multidisciplinary teamwork, selecting candidates for DBS, managing expectations of families, and measuring clinical improvements. The thematic codes that I developed were oriented around these challenges, and I was able to use a feedback session with the team to check that the challenges I had identified were in fact representative of their work. In this way, I settled upon four major challenges in the establishment of the PMDS proto-platform: Coordinating multidisciplinary (Chapter 4); selecting candidates for deep brain stimulation (Chapter 5); managing expectations of families (Chapter 6), and measuring clinical improvements (Chapter 7).

3. A History of Deep Brain Stimulation

This chapter is based on information gathered from a range of resources. Scientific papers from 1930 onwards relating to electrostimulation, neurostimulation, neuromodulation, and deep brain stimulation were sourced from a range of neurosurgical, neurological, psychiatric journals, as well as engineering and bioengineering publications. Data were also gathered from scientific papers relating to the development of clinical assessment tools for movement disorders published from the 1970s onwards. Newspaper articles, press releases, transcripts of FDA panel hearings and meetings (available online), and secondary historical documents, and accounts produced by engineers, medical advisors, and numerous clinicians, particularly the neurosurgeons Philip Gildenberg (Gildenberg 2005, 2009, 2000) and Adrian Upton (Upton 1986; Upton and Lazorthes 1987) were also included.

Bibliography

4 News. 2012. "Revealed: Postcode Lottery for Dystonia Treatment." accessed 29 June. http://www.channel4.com/news/revealed-the-postcode-lottery-for-dystonia-treatment.

Abrishami, Payam, Albert Boer, and Klasien Horstman. 2014. "Understanding the Adoption Dynamics of Medical Innovations: Affordances of the Da Vinci Robot in the Netherlands." *Social Science & Medicine* 117(0): 125–133. doi: http://dx.doi.org/10.1016/j.socscimed.2014.07.046.

Ackerman, S. 2006. *Hard Science, Hard Choices: Facts, Ethics, and Policies Guiding Brain Science Today*. New York: Dana Press.

Adams, A. 2008. *Medicine by Design: The Architect and the Modern Hospital, 1893–1943*. Minneapolis: University of Minnesota Press.

Agid, Y., M. Schüpbach, M. Gargiulo, L. Mallet, J. L. Houeto, C. Behar, D. Maltête, V. Mesnage, and M. L. Welter. 2006. "Neurosurgery in Parkinson's Disease: The Doctor is Happy, the Patient Less So?" In *Parkinson's Disease and Related Disorders*, edited by P. Riederer, H. Reichmann, M. B. H. Youdim, and M. Gerlach, 409–414. Vienna: Springer.

Akrich, M. 1992. "The De-Scription of Technological Objects." In *Shaping Technology /Building Society: Studies in Socio-Technical Change*, edited by W. Bijker and J. Law, 204–224. London: MIT University Press.

© The Author(s) 2017   **219**
J. Gardner, *Rethinking the Clinical Gaze*, Health, Technology and Society, DOI 10.1007/978-3-319-53270-7

Akrich, M, and B Latour. 1992. "A Summary of Convenient Vocabulary for the Semiotics of Human Interaction and Nonhuman Assemblies." In *Shaping Technology, Building Society*, edited by W. Bijker and J. Law, 259–264. Cambridge MA: MIT Press.

Alcadipani, Rafael, and John Hassard. 2010. "Actor-Network Theory, Organizations and Critique: Towards a Politics of Organizing." *Organization* 17(4): 419–435. doi: 10.1177/1350508410364441.

Andy, O. J. 1983. "Thalamic Stimulation for Control of Movement Disorders." *Appl Neurophysiol* 46(1-4): 107–111.

Antonelli, Cristiano. 2009. "The Economics of Innovation: From the Classical Legacies to the Economics of Complexity." *Economics of Innovation and New Technology* 18(7): 611–646. doi: 10.1080/10438590802564543.

Antonelli, Cristiano, and Gianluigi Ferraris. 2011. "Innovation as an Emerging System Property: An Agent Based Simulation Model." *Journal of Artificial Societies and Social Simulation* 14(2): 1. doi: 10.18564/jasss.1741.

Armstrong, D., R. Lilford, J. Ogden, and S. Wessely. 2007. "Health-Related Quality of Life and the Transformation of Symptoms." *Sociol Health Illn* 29(4): 570–583. doi: 10.1111/j.1467-9566.2007.01006.x.

Austin, Allana. 2015. "Parental Experiences of Secondary Dystonia and the Journey Through Deep Brain Stimulation." Doctor in Clinical Pscyhology, Royal Holloway.

Austin, Allana, Jean-Pierre Lin, Richard Selway, Keyoumars Ashkan, and Tamsin Owen. 2016. "What Parents Think and Feel About Deep Brain Stimulation in Paediatric Secondary Dystonia Including Cerebral Palsy: A Qualitative Study of Parental Decision-Making." *European Journal of Paediatric Neurology*. doi: 10.1016/j.ejpn.2016.08.011.

Baer, L., D. R. Elford, and P. Cukor. 1997. "Telepsychiatry at Forty: What Have We Learned?" *Harv Rev Psychiatry* 5(1): 7–17.

Barry, A. 2001. *Political Machines:Governing a Technological Society*. New York: Athlone Press.

Barry, Andrew, Georgina Born, and Gisa Weszkalnys. 2008. "Logics of Interdisciplinarity." *Economy and Society* 37(1): 20–49. doi: 10.1080/03085140701760841.

Bell, Emily, Ghislaine Mathieu, and Eric Racine. 2009. "Preparing the Ethical Future of Deep Brain Stimulation." *Surgical Neurology* 72(6): 577–586.

Bell, K. 2013. "Biomarkers, the Molecular Gaze and the Transformation of Cancer Survivorship." *Biosocieties* 8(2): 124–143. doi: 10.1057/biosoc.2013.6.

Benabid, A. L., P. Pollak, A. Louveau, S. Henry, and J. De Rougemont. 1987. "Combined (Thalamotomy and Stimulation) Stereotactic Surgery of the VIM Thalamic Nucleus for Bilateral Parkinson Disease." *Appl Neurophysiol* 50(1-6): 344–346.

Benabid, A. L., P. Pollak, D. Hoffmann, C. Gervason, M. Hommel, J. E. Perret, J. De Rougemont, and D. M. Gao. 1991. "Long-Term Suppression of Tremor by Chronic Stimulation of the Ventral Intermediate Thalamic Nucleus." *The Lancet* 337(8738): 403–406. doi: 10.1016/0140-6736(91)91175-t.

Berg, M. 1998. "Order(s) and Disorder(s): Of Protocols and Medical Practices." In *Differences in Medicine: Unraveling Practices, Techniques and Bodies*, edited by M. Berg and Annemarie Mol, 227–246. Durham: Duke University Press.

Bergman, H., T. Wichmann, and M.R. DeLong. 1990. "Reversal of Experimental Parkinsonism by Lesions of the Subthalamic Nucleus." *Science* 249(4975): 1436–1438. doi: 10.1126/science.2402638.

Bijker, W.E., T.P. Hughes, and T.J. Pinch. 1987. *The Social Construction of Technological Systems:New Directions in the Sociology and History of Technology*. London: MIT Press.

Blomstedt, Patric, and Marwan I. Hariz. 2010. "Deep Brain Stimulation for Movement Disorders Before DBS for Movement Disorders." *Parkinsonism & Related Disorders* 16(7): 429–433. doi: 10.1016/j.parkreldis.2010.04.005.

Blume, Stuart. 2010. *The Artificial Ear: Cochlear Implants and the Culture of Deafness*. London: Rutgers University Press.

Bodewitz, Henk, Henk Buurma, and Gerard de Vries.. 1989. "Regulatory Science and the Social Management of Trust in Medicine." In *The Social Construction of Technological Systems: New Directions in the Sociology and History of Technology*, edited by Wiebe Bijker, Thomas Hughes, and Trevor Pinch, 241–259. London: MIT Press.

Bolton, Sharon. 2001. "Changing Faces: Nurses as Emotional Jugglers." *Sociology of Health & Illness* 23(1): 85–100. doi: 10.1111/1467-9566.00242.

Boon, Paul, Kristl Vonck, Veerle De Herdt, Annelies Van Dycke, Maarten Goethals, Lut Goossens, Michel Van Zandijcke, Tim De Smedt, Isabelle Dewaele, Rik Achten, Wytse Wadman, Frank Dewaele, Jacques Caemaert, and Van Roost. Dirk. 2007. "Deep Brain Stimulation in Patients with Refractory Temporal Lobe Epilepsy." *Epilepsia* 48(8): 1551–1560. doi: 10.1111/j.1528-1167.2007.01005.x.

Borup, Mads, Nik Brown, Kornelia Konrad, and Harro Van Lente. 2006. "The Sociology of Expectations in Science and Technology." *Technology Analysis & Strategic Management* 18(3-4): 285–298. doi: 10.1080/09537320600777002.

Bourret, Pascale. 2005. "BRCA Patients and Clinical Collectives: New Configurations of Action in Cancer Genetics Practices." *Social Studies of Science* 35(1): 41–68. doi: 10.1177/0306312705048716.

Breggin, M. 1972. "The Return of Lobotomy and Psychosurgery." In *Congressional Record*, Washington DC.: United States Government.

Brice, J., and L. McLellan. 1980. "Suppression of Intention Tremor by Contingent Deep-Brain Stimulation." *Lancet* 1(8180): 1221–1222.

Broer, Tineke, and Martyn Pickersgill. 2015a. "(Low) Expectations, Legitimization, and the Contingent Uses of Scientific Knowledge: Engagements with Neuroscience in Scottish Social Policy and Services." *2015* 1:20.

Broer, Tineke, and Martyn Pickersgill. 2015b. "Targeting Brains, Producing Responsibilities: The Use of Neuroscience within British Social Policy." *Social Science & Medicine* 132: 54–61. doi: http://dx.doi.org/10.1016/j.socscimed.2015.03.022.

Brown, N., B. Rappert, and A. Webster. 2000. *Contested Futures: A Sociology of Prospective Techno-Science*. Farnham: Ashgate.

Brown, Nik, and Mike Michael. 2003. "A Sociology of Expectations: Retrospecting Prospects and Prospecting Retrospects." *Technology Analysis & Strategic Management* 15(1): 3–18. doi: 10.1080/0953732032000046024.

Brown, Nik, and Andrew Webster. 2004. *New Medical Technologies and Society: Reordering Life*. Cambridge: Polity.

Brown, P. R., A. Alaszewski, T. Swift, and A. Nordin. 2011. "Actions Speak Louder than Words: The Embodiment of Trust by Healthcare Professionals in Gynae-Oncology." *Sociol Health Illn* 33(2): 280–295. doi: 10.1111/j.1467-9566.2010.01284.x.

Burke, D., J. D. Gillies, and J. W. Lance. 1970. "The Quadriceps Stretch Reflex in Human Spasticity." *J Neurol Neurosurg Psychiatry* 33(2): 216–223.

Burke, R. E., S. Fahn, C. D. Marsden, S. B. Bressman, C. Moskowitz, and J. Friedman. 1985. "Validity and Reliability of a Rating Scale for the Primary Torsion Dystonias." *Neurology* 35(1): 73–77.

Buskens, E., and J. Van Gijn. 2001. "Ethics, Outcome Variables and Clinical Scales: The Clinician's Point of View." In *Clinical Trials in Neurology*, edited by Roberto J. Guiloff, 29–41. London: Springer London.

Carmel, Simon. 2013. "The Craft of Intensive Care Medicine." *Sociology of Health & Illness* 35(5): 731–745. doi: 10.1111/j.1467-9566.2012.01524.x.

Centellas, Kate M., Regina E. Smardon, and Steve Fifield. 2014. "Calibrating Translational Cancer Research: Collaboration Without Consensus in Interdisciplinary Laboratory Meetings." *Science, Technology & Human Values* 39(3): 311–335. doi: 10.1177/0162243913505650.

Chou, K. L., S. Grube, and P.G. Patil. 2011. *Deep Brain Stimulation: A New Life for People with Parkinson's, Dystonia,and Essential Tremor.* New York: Demos Medical Publishing.

Clarke, Adele E., Janet K. Shim, Laura Mamo, Jennifer Ruth Fosket, and Jennifer R. Fishman. 2003. "Biomedicalization: Technoscientific Transformations of Health, Illness, and U.S. Biomedicine." *American Sociological Review* 68(2): 161–194. doi: 10.2307/1519765.

Clarke, A.E., L. Mamo, J.R. Fosket, J.R. Fishman, J.K. Shim, and E. Riska. 2009. *Biomedicalization: Technoscience, Health, and Illness in the U.S.* Durham NC: Duke University Press.

Coffey, R. 2001. "Deep Brain Stimulation for Chronic Pain: Results of Two Multicenter Trials and a Structured Review." *Pain Medicine* 2(3): 183–192. doi: 10.1046/j.1526-4637.2001.01029.x.

Coffey, Robert. 2009. "Deep Brain Stimulation Devices: A Brief Technical History and Review." *Artificial Organs* 33(3): 208–220. doi: 10.1111/j.1525-1594.2008.00620.x.

Coffey, Robert, and Andres Lozano. 2006. "Neurostimulation for Chronic Noncancer Pain: An Evaluation of the Clinical Evidence and Recommendations for Future Trial Designs." *Journal of Neurosurgery* 105: 175–189.

Consoli, Davide, and Andrea Mina. 2008. "An Evolutionary Perspective on Health Innovation Systems." *Journal of Evolutionary Economics* 19(2): 297–319. doi: 10.1007/s00191-008-0127-3.

Coombs, M. 2003. "Power and Conflict in Intensive Care Clinical Decision Making." *Intensive Crit Care Nurs* 19(3): 125–135.

Coombs, M., and S. J. Ersser. 2004. "Medical Hegemony in Decision-Making–a Barrier to Interdisciplinary Working in Intensive Care?" *J Adv Nurs* 46(3): 245–252. doi: 10.1111/j.1365-2648.2004.02984.x.

Cooper, I., and A. Upton. 1978. "Use of Chronic Cerebellar Stimulation for Disorders of Disinhibition." *The Lancet* 311(8064): 595–600. doi: 10.1016/s0140-6736(78)91038-3.

Cooper, I., A. Upton, and I. Amin. 1980. "Reversibility of Chronic Neurologic Deficits. Some Effects of Electrical Stimulation of the Thalamus and Internal Capsule in Man." *Applied Neurophysiology* 43: 244–258.

Cooper, I., A. Upton, and I. Amin. 1982. "Chronic Cerebellar Stimulation (CCS) and Deep Brain Stimulation (DBS) in Involuntary Movement Disorders." *Applied Neurophysiology* 45: 207–217.

Corea, G. 1985. *The Mother Machine: Reproductive Technologies from Artificial Insemination to Artificial Wombs.* New York: Harper & Row.

Cribb, A. 2011. *Involvement, Shared Decision-Making and Medicines.* London: Royal Pharmaceutical Society.

Crossley, Nick. 2001. "The Phenomenological Habitus and Its Construction." *Theory and Society* 30(1): 81–120.

D'Amour, D., M. Ferrada-Videla, L. San Martin Rodriguez, and M. D. Beaulieu. 2005. "The Conceptual Basis for Interprofessional Collaboration: Core Concepts and Theoretical Frameworks." *J Interprof Care* 19 Suppl 1: 116–131. doi: 10.1080/13561820500082529.

Danish, S., and G. Baltuch. 2007. "History of Deep Brain Stimulation." In *Deep Brain Stimulation for Parkinson's Disease*, edited by G. Baltuch and M. Stern, 1–16. Boca Raton FL: Taylor & Francis.

De Laat, Bastiaan. 2000. "Scripts for the Future: Using Innovation Studies to Design Foresight Tools." In *Contested Futures*, edited by N. Brown, B. Rappert, and A. Webster, 175–208. Farnham: Ashgate.

Defazio, G. 2010. "The Epidemiology of Primary Dystonia: Current Evidence and Perspectives." *Eur J Neurol* 17 Suppl 1: 9–14. doi: 10.1111/j.1468-1331.2010.03053.x.

DeLong, Mahlon R. 1990. "Primate Models of Movement Disorders of Basal Ganglia Origin." *Trends in Neurosciences* 13(7): 281–285. doi: http://dx.doi.org/10.1016/0166-2236(90)90110-V.

Department of Health. 1999. *Patient and Public Involvement in the new NHS.* London: The Stationery Office.

Department of Health. 2000. *The NHS Plan: A Plan for Investment, A Plan for Reform.* London: The Crown.

Department of Health. 2003. *Building on the Best: Choice, Responsiveness and Equity in the NHS.* London: The Stationary Office.

Department of Health. 2008. *NHS Next Stage Review: A High Qaulity Workforce.* London: The Crown.

Department of Health. 2012. *2012-2013 Tariff Information Spreadsheet.* London: The Crown.

Deuschl, Günther, Carmen Schade-Brittinger, Paul Krack, Jens Volkmann, Helmut Schäfer, Kai Bötzel, Christine Daniels, Angela Deutschländer, Ulrich Dillmann, Wilhelm Eisner, Doreen Gruber, Wolfgang Hamel, Jan

Herzog, Rüdiger Hilker, Stephan Klebe, Manja Kloß, Jan Koy, Martin Krause, Andreas Kupsch, Delia Lorenz, H. Stefan Lorenzl, Maximilian Mehdorn, Jean Richard Moringlane, Marcus O. Wolfgang Oertel, Heinz Reichmann Pinsker, Alexander Reuß, Gerd-Helge Schneider, Alfons Schnitzler, Ulrich Steude, Volker Sturm, Lars Timmermann, Volker Tronnier, Thomas Trottenberg, Lars Wojtecki, Elisabeth Wolf, Werner Poewe, and Jürgen Voges. 2006. "A Randomized Trial of Deep-Brain Stimulation for Parkinson's Disease." *New England Journal of Medicine* 355(9): 896–908. doi: 10.1056/NEJMoa060281.

Djellal, Faridah, and Faïz Gallouj. 2005. "Mapping Innovation Dynamics in Hospitals." *Research Policy* 34(6): 817–835. doi: http://dx.doi.org/10.1016/j.respol.2005.04.007.

DMRF. 2010. "Deep Brain Foundation." Dystonia Medical Research Foundation, accessed 2013. http://www.dystonia-foundation.org/pages/deep_brain_stimulation/151.php.

DoH. 2003. *Building on the Best: Choice, Responsiveness and Equity in the NHS.* London: Department of Health.

DoH. 2013. *Integrated Care and Support: Our Shared Commitment.* London: Department of Health.

Dubbin, L. A., J. S. Chang, and J. K. Shim. 2013. "Cultural Health Capital and the Interactional Dynamics of Patient-Centered Care." *Soc Sci Med* 93: 113–120. doi: 10.1016/j.socscimed.2013.06.014.

Dumit, Joseph. 2004. *Picturing Personhood: Brain Scans and Biomedical Identity.* New Jersey: Princeton.

The Dystonia Society. 2010. "Deep Brain Stimulation: The Campain Continues." The Dystonia Society, accessed 29 June. http://www.dystonia.org.uk/index.php/7-about-us/about-us/266-deep-brain-stimulation-the-campaign-continues.

Dystonia Society, UK. 2014. *A Guide to Best Practice for Health and Social Care Professionals.* London: The Dystonia Society UK.

East Midlands Specialised Commissioning Group. 2011. "A Commissioning Policy for DBS." accessed 1 July. http://www.emscg.nhs.uk/Library/EMSCGPolicyforDBSV1D6.pdf.

Epstein, Ronald M., and Richard L. Street. 2011. "The Values and Value of Patient-Centered Care." *The Annals of Family Medicine* 9(2): 100–103. doi: 10.1370/afm.1239.

Fahn, S., and R. Elton. 1987. "Unified Parkinson's Disease Rating Scale (UPDRS)." In *Recent Developments in Parkinson's Disease*, edited by S. Fahn,

C. D. Marsden, D. B. Calne, and M. Goldstein, 153–163, 293–304. Florham Park: Macmillan Health Care Information.

Faulkner, Alex. 2009. *Medical Technology into Healthcare and Society: A Sociology of Devices*. Basingstoke: Palgrave MacMillan.

FDA. 1997. "Premarket Notification: Implantable Deep Brain Stimulation for the Treatment of Tremor due to Parkinson's Disease and Essential Tremor." In *Neurological Devices Panel of the Medical Devices Advisory Committee*. Washington D.C.: Food and Drug Administration.

Fein, E. 2011. "Innocent Machines: Asperger's Syndrome and the Neurostructural Self." In *Sociological Reflections on the Neurosciences*, M. Pickersgill and I. Van Keulen, 27–49. Bingley: Emerald Group Publishing Limited.

Finn, R., M. Learmonth, and P. Reedy. 2010. "Some Unintended Effects of Teamwork in Healthcare." *Soc Sci Med* 70(8): 1148–1154. doi: 10.1016/j.socscimed.2009.12.025.

Fins, J. J., and N. D. Schiff. 2010. "Conflicts of Interest in Deep Brain Stimulation Research and the Ethics of Transparency." *J Clin Ethics* 21(2): 125–132.

Fins, J. J., H. S. Mayberg, B. Nuttin, C. S. Kubu, T. Galert, V. Sturm, K. Stoppenbrink, R. Merkel, and T. E. Schlaepfer. 2011. "Misuse of the FDA's Humanitarian Device Exemption in Deep Brain Stimulation for Obsessive-Compulsive Disorder." *Health Aff (Millwood)* 30(2): 302–311. doi: 10.1377/hlthaff.2010.0157.

Fisher, Anne. 2001. *Assessment of Motor and Process Skills: User Manual*. Fort Collins, CO: Three Star Press, Incorporated.

Fisher, Anne. 2003. *Assessment of Motor and Process Skills*. 6th *edn*. Fort Collins, CO: Fort Collins, CO.

Fisher, A. G., and K. B. Jones. 2010. *Assessment of Motor and Process Skills. Vol. 1: Development, Standardization, and Administration Manual*. 7 ed. Fort Collins: Three Star Press.

Fitzgerald, Des. 2014. "The Trouble with Brain Imaging: Hope, Uncertainty and Ambivalence in the Neuroscience of Autism." *BioSocieties* 9(3): 241–261. doi: 10.1057/biosoc.2014.15.

Foote, Susan Bartlett. 1978. "Loops and Loopholes: Hazardous Device Regulation Under the 1976 Medical Device Amendments to the Food, Drug and Cosmetic Act." *Ecology Law Quarterly* 7: 101–135.

Foucault, Michel. 1991. *Discipline and Punish: The Birth of the Prison*. London: Penguin.

Foucault, Michel. 2003. *The Birth of the Clinic*. London: Routledge Classics.

Fournier, Valerie. 2000. "Boundary Making and the (un)making of the Professions." In *Professionalism, Boundaries and the Workplace*, edited by N. Malin, 67–86. London: Routledge.

Fox, N. J. 2012. *The Body*: Wiley.

Fox, Nick J. 2013. "Creativity and Health: An Anti-Humanist Reflection." *Health* 17(5): 495–511. doi: 10.1177/1363459312464074.

Fraix, V., J.-L. Houeto, C. Lagrange, C. Le Pen, P. Krystkowiak, D. Guehl, C. Ardouin, M.-L. Welter, F. Maurel, L. Defebvre, A. Rougier, A.-L. Benabid, V. Mesnage, M. Ligier, S. Blond, P. Burbaud, B. Bioulac, A. Destée, P. Cornu, and P. Pollak. 2006. "Clinical and Economic Results of Bilateral Subthalamic Nucleus Stimulation in Parkinson's Disease." *Journal of Neurology, Neurosurgery & Psychiatry* 77(4): 443–449. doi: 10.1136/jnnp.2005.077677.

Gardner, John. 2013. "A History of Deep Brain Stimulation: Technological Innovation and the Role of Clinical Assessment Tools." *Social Studies of Science* 43(5): 707–728. doi: 10.1177/0306312713483678.

Gardner, J. 2016a. "Securing a Future for Responsible Neuromodulation in Children: The Importance of Maintaining a Broad Clinical gaze." *Eur J Paediatr Neurol*. doi: 10.1016/j.ejpn.2016.04.019.

Gardner, John. 2016b. "Patient-Centred Medicine and the Broad Clinical Gaze: Measuring Outcomes in Paediatric Deep Brain Stimulation." *BioSocieties*. doi: 10.1057/biosoc.2016.6.

Gardner, John, and Alan Cribb. 2016. "The Dispositions of Things: The Non-Human Dimension of Power and Ethics in Patient-Centred Medicine." *Sociology of Health & Illness* 38(7): 1043–1057. doi: 10.1111/1467-9566.12431.

Gardner, John, and Andrew Webster. 2016. "The Social Management of Biomedical Novelty: Facilitating Translation in Regenerative Medicine." *Social Science & Medicine* 156: 90–97. doi: http://dx.doi.org/10.1016/j.socscimed.2016.03.025.

Gardner, John, and Clare Williams. 2015. "Corporal Diagnostic Work and Diagnostic Spaces: Clinicians' Use of Space and Bodies During Diagnosis." *Sociology of Health & Illness* 37(5): 765–781. doi: 10.1111/1467-9566.12233.

Gardner, John, Gabrielle Samuel, and Clare Williams. 2015. "Sociology of Low Expectations: Recalibration as Innovation Work in Biomedicine." *Science, Technology & Human Values* 40(6): 998–1021. doi: 10.1177/0162243915585579.

Gelijns, A., and N. Rosenberg. 1994. "The Dynamics of Technological Change in Medicine." *Health Aff (Millwood)* 13(3): 28–46.

Geyer, H. L., and S. B. Bressman. 2006. "The Diagnosis of Dystonia." *Lancet Neurol* 5(9): 780–790. doi: 10.1016/s1474-4422(06)70547-6.

Gieryn, Thomas F. 2002. "What Buildings Do." *Theory and Society* 31(1): 35–74.

Gilbert, Frédéric, and Daniela Ovadia. 2011. "Deep Brain Stimulation in the Media: Over-Optimistic Portrayals Call for a New Strategy Involving Journalists and Scientists in Ethical Debates." *Frontiers in Integrative Neuroscience* 5. doi: 10.3389/fnint.2011.00016.

Gildenberg, P. 2000. "Fifty Years of Stereotactic and Functional Neurosurgery." In *Fifty Years of Neurosurgery*, edited by D. Barrow, D Kondziolka, E. Laws, and V Traynelis, 295–320. New York: Lippincott Williams & Wilkins.

Gildenberg, P. 2005. "Evolution of Neuromodulation." *Stereotactic and Functional Neurosurgery* 83(2-3): 71–79.

Gildenberg, P. 2009. "Neuromodulation: A Historical Perspective." In *Neuromodulation*, edited by E. Krames, P. Peckham, and A. Rezai, 9–20. Amsterdam: Elsevier & Academic Press.

Gimeno, H, and Jean-Pierre Lin. 2016. "The International Classification of Functioning (ICF) to Evaluate Deep Brain Stimulation Neuromodulation in Childhood Dystonia-Hyperkinesia Informs Future Clinical & Research Priorities in a Multidisciplinary Model of Care." *European Journal of Paediatric Neurology* 21(1): 147–167.

Gimeno, H., K. Tustin, R. Selway, and J. P. Lin. 2012. "Beyond the Burke-Fahn-Marsden Dystonia Rating Scale: Deep Brain Stimulation in Childhood Secondary Dystonia." *Eur J Paediatr Neurol* 16(5): 501–508. doi: 10.1016/j.ejpn.2011.12.014.

Gisquet, Elsa. 2008. "Cerebral Implants and Parkinson's Disease: A Unique Form of Biographical Disruption?" *Social Science & Medicine* 67(11): 1847–1851.

Gliklich, R., N. Dreyer, and M. Leavey. eds. 2014. *Registries for Evaluating Patient Outcomes: A User's Guide, Agency for Healthcare Research and Quality.* Rockville, MD: AHRQ.

Goetz, C., Glenn T. Stebbins, Teresa A. Chmura, Stanley Fahn, Harold L. Klawans, and C. David Marsden. 1995. "Teaching Tape for the Motor Section of the Unified Parkinson's Disease Rating Scale." *Movement Disorders* 10(3): 263–266. doi: 10.1002/mds.870100305.

Goetz, C., O. Werner Poewe, Rascol C. Sampaio, G. Stebbins, S. Fahn, A. E. Lang, P. Martinez-Martin, B. Tilley, and Bob Johannes van Hilten. 2003. "The Unified Parkinson's Disease Rating Scale (UPDRS): Status and

Recommendations." *Movement Disorders* 18(7): 738–750. doi: 10.1002/mds.10473.

Goodwin, Charles. 1994. "Professional Vision." *American Anthropologist* 96(3): 606–633.

Goodwin, D. 2010. "Sensing the Way: Embodied Dimension of Diagnostic Work." In *Ethnographies of Diagnostic Work: Dimensions of Transformative Practice*, edited by M. Buscher, D. Goodwin, and J. Mesman, 73–94. Basingstoke: Palgrave MacMillan.

Gordon, L. M., J. L. Keller, E. E. Stashinko, A. H. Hoon, and A. J. Bastian. 2006. "Can Spasticity and Dystonia be Independently Measured in Cerebral Palsy?" *Pediatr Neurol* 35(6): 375–381. doi: 10.1016/j.pediatrneurol.2006.06.015.

Goto, S., A. G. Fisher, and W. L. Mayberry. 1996. "The Assessment of Motor and Process Skills Applied Cross-Culturally to the Japanese." *Am J Occup Ther* 50(10): 798–806.

Greenhalgh, T., G. Robert, F. Macfarlane, P. Bate, and O. Kyriakidou. 2004. "Diffusion of Innovations in Service Organizations: Systematic Review and Recommendations. *Milbank Q* 82. doi: 10.1111/j.0887-378X.2004.00325.x.

Gregory, Allison, and Susan Hayflick. 2005. "Neurodegeneration with Brain Iron Accumulation." *Folia Neuropathologica* 43(4): 286–296.

Grosz, Elizabeth. 2005. "Bergson, Deleuze and the Becoming of Unbecoming." *Parallax* 11(2): 4–13. doi: 10.1080/13534640500058434.

Guggenheim, Michael. 2012. "Laboratizing and De-Laboratizing the World: Changing Sociological Concepts for Places of Knowledge Production." *History of the Human Sciences* 25(1): 99–118. doi: 10.1177/0952695111422978.

Hammersley, M., and P. Atkinson. 2007. *Ethnography:Principles in Practice*. Abingdon: Taylor & Francis.

Haraway, Donna. 1988. "Situated Knowledges: The Science Question in Feminism and the Privilege of Partial Perspective." *Feminist Studies* 14(3): 575–599. doi: 10.2307/3178066.

Hardt, M. 1995. *Gilles Deleuze: An Apprenticeship in Philosophy*. Minneapolis: University of Minnesota Press.

Hardy, Brian, and Adrian Turrell. 1992. *Innovations in Community Care Management: Minimising Vulnerability*. Aldershot: Avebury.

Hariz, M., Patric Blomstedt, and L. Zrinzo. 2010. "Deep Brain Stimulation Between 1947 and 1987: The Untold Story." *Neurosurgical Focus* 29(2): E1.

Harris, Anna. 2011. "In a Moment of Mismatch: Overseas Doctors' Adjustments in new Hospital Environments." *Sociology of Health & Illness* 33(2): 308–320. doi: 10.1111/j.1467-9566.2010.01307.x.

Haynes, R. Brian. 1990. "Loose Connections Between Peer-Reviewed Clinical Journals and Clinical Practice." *Annals of Internal Medicine* 113(9): 724–728. doi: 10.7326/0003-4819-113-9-724.

Heath, C. 1986. *Body Movement and Speech in Medical Interaction*. Cambridge MA: Cambridge University Press.

Heath, Robert. 1977. "Modulation of Emotion with a Brain Pacemaker: Treatment for Intractable Psychiatric Illness." *Journal of Nervous and Mental Disease* 165: 300–317.

Heath, Christian. 2002. "Demonstrative Suffering: The Gestural (Re)embodiment of Symptoms." *Journal of Communication* 52(3): 597–616. doi: 10.1111/j.1460-2466.2002.tb02564.x.

Heath, R. G., R. C. Llewellyn, and A. M. Rouchell. 1980. "The Cerebellar Pacemaker for Intractable Behavioral Disorders and Epilepsy: Follow-Up Report." *Biol Psychiatry* 15(2): 243–256.

Henke, C., and Thomas F. Gieryn. 2008. "Sites of Scientific Practice: The Enduring Importance of Place." In *The Handbook of Science and Technology Studies*, edited by E. Hackett, O. Amsterdamska, M. Lynch, and Wajcman, 353–376. Cambridge MA: MIT Press.

Hester, Micah. 2007. "Guest Editorial: Technology and the Body." *Cambridge Quarterly of Healthcare Ethics* 16(03): 254–256. doi: 10.1017/S0963180107070284.

Hill, Benjamin. 2010. *The Sociology of Innovation*. Cambridge MA: MIT.

Hobart, J. C., and A. J. Thompson. 2001. "Assessment Measures and Clinical Scales." In *Clinical Trials in Neurology*, edited by Roberto J. Guiloff, 17–28. London: Springer London.

Hopkins, Michael M., Paul A. Martin, Paul Nightingale, Alison Kraft, and Surya Mahdi. 2007. "The Myth of the Biotech Revolution: An Assessment of Technological, Clinical and Organisational Change." *Research Policy* 36(4): 566–589. doi: http://dx.doi.org/10.1016/j.respol.2007.02.013.

Hosobuchi, Y., J. E. Adams, and B. Rutkin. 1973. "Chronic Thalamic Stimulation for the Control of Facial Anesthesia Dolorosa." *Archives of Neurology* 29(3): 158–161.

House of Lords Science and Technology Committee. 2013. *Regenerative Medicine Report*. London: House of Lords.

Housley, W. 2003. *Interaction in Multidisciplinary Teams*. Farnham: Ashgate.

Jankovic, J. 2007. "Dystonic Disorders." In *Parkinson's Disease and Movement Disorders*, edited by J. Jankovic and E. Tolosa, 321–347. Philadelphia: Lippincott Williams & Wilkins.

Jasanoff, S. 2004. "Ordering Knowledge, Ordering Society." In *States of Knowledge: The Co-Production of Science and the Social Order*, edited by S. Jasanoff, 13–45. London: Routledge.

Jespersen, Astrid P., Julie Bønnelycke, and Hanne H. Eriksen. 2014. "Careful Science? Bodywork and Care Practices in Randomised Clinical Trials." *Sociology of Health & Illness* 36(5): 655–669. doi: 10.1111/1467-9566.12094.

Joyce, Kelly A. 2006. "From Numbers to Pictures: The Development of Magnetic Resonance Imaging and the Visual Turn in Medicine." *Science as Culture* 15(1): 1–22. doi: 10.1080/09505430600639322.

Joyce, Kelly. 2008. *Magnetic Appeal: MRI and the Myth of Transperancy*. London: Cornell University Press.

Keating, P., and A. Cambrosio. 2003. *Biomedical Platforms: Realigning the Normal and the Pathological in Late-Twentieth-Century Medicine*. Cambridge MA: MIT Press.

Kennedy, Ian. 2001. *The Report of the Public Inquiry into Children's Heart Surgery at the Bristol Royal Infirmary 1984-1995: Learning from Bristol*. London: The Crown.

Klein, R., J. G. Raymond, and L. J. Dumble. 2013. *Ru 486: Misconceptions, Myths and Morals*. Melbourne: Spinifex Press.

Krystkowiak, P., S. T. Du Montcel, L. Vercueil, J. L. Houeto, C. Lagrange, P. Cornu, S. Blond, A. L. Benabid, P. Pollak, and M. Vidailhet. 2007. "Reliability of the Burke-Fahn-Marsden Scale in a Multicenter Trial for Dystonia." *Mov Disord* 22(5): 685–689. doi: 10.1002/mds.21392.

La Chapelle, C., F. Jansen, B. Pelger, and B. Mol. 2013. ""Robotchirurgie in Nederland: Hoogwaardig Bewijs Voor Effectiviteit Ontbreekt." *Ned. Tijdschr. Geneeskd* 157: A515.

Laine, C., and F. Davidoff. 1996. "Patient-Centered Medicine. A Professional Evolution." *Jama* 275(2): 152–156.

Laitinen, L. V., A. T. Bergenheim, and M. I. Hariz. 1992. "Leksell's Posteroventral Pallidotomy in the Treatment of Parkinson's Disease." *Journal of Neurosurgery* 76(1): 53–61.

Latimer, J. 1997. "Giving Patients a Future: The Constituting of Classes in an Acute Medical Unit." *Sociology of Health & Illness* 19(2): 160–185. doi: 10.1111/1467-9566.ep10934396.

Latimer, J. 2014. "Dwelling with Dementia: Body-Self Relations, Participation and Care." In *Symposium on Researching (Bio)Medicine with Carecosmopolitics, Affects and Ethics*, edited by J. Latimer, M. Schillmer, and A. Kerr. University of York: BSA Medical Sociology Study Group Annual Conference.

Latour, B. 1987. *Science in Action: How to Follow Scientists and Engineers Through Society*. Cambridge MA: Harvard University Press.

Latour, B. 1990. "Drawing Things Together." In *Representations of Scientific Practice*, edited by M. Lynch and S. Woolgar, 19–68. Cambridge MA: MIT Press.

Latour, B. 1992. "Where Are the Missing Masses? The Sociology of a Few Mundane Artifacts." In *Shaping Technology, Building Society: Studies in Sociotechnical Change*, edited by W. E. Bijker and J. Law, 225–258. Cambridge MA: MIT Press.

Latour, Bruno. 2004. "How to Talk About the Body? the Normative Dimension of Science Studies." *Body & Society* 10(2-3): 205–229. doi: 10.1177/1357034x04042943.

Latour, Bruno. 2005. *Reassembling the Social: An Introduction to Actor-Network Theory*. Oxford: Oxford University Press.

Latour, B., and S. Woolgar. 1979. *Laboratory Life: The Construction of Scientific Facts*. Princeton: Princeton University Press.

Law, John. 2009. "Actor Network Theory and Material Semiotics." In *The New Blackwell Companion to Social Theory*, edited by B. Turner, 141–158. Hoboken NJ: Wiley-Blackwell.

Law, M., S. Babtiste, A. Carswell, M. McColl, H. Polotajko, and N. Pollock. 2005. *Canadian Occupational Performance Measure*. Ottowa: COAT Publications ACE.

Lebiedowska, M. K., D. Gaebler-Spira, R. S. Burns, and J. R. Fisk. 2004. "Biomechanic Characteristics of Patients with Spastic and Dystonic Hypertonia in Cerebral Palsy." *Arch Phys Med Rehabil* 85(6): 875–880.

Levenstein, J. H., E. C. McCracken, I. R. McWhinney, M. A. Stewart, and J. B. Brown. 1986. "The Patient-Centred Clinical Method. 1. A Model for the Doctor-Patient Interaction in Family Medicine." *Fam Pract* 3(1): 24–30.

Liberati, Elisa Giulia, Mara Gorli, Lorenzo Moja, Laura Galuppo, Silvio Ripamonti, and Giuseppe Scaratti. 2015. "Exploring the Practice of Patient Centered Care: The Role of Ethnography and Reflexivity." *Social Science & Medicine* 133: 45–52. doi: http://dx.doi.org/10.1016/j.socscimed.2015.03.050.

Limousin, P., P. Pollak, A. Benazzouz, D. Hoffmann, J. F. Le Bas, J. E. Perret, A. L. Benabid, and El Broussolle. 1995. "Effect on Parkinsonian Signs and Symptoms of Bilateral Subthalamic Nucleus Stimulation." *The Lancet* 345(8942): 91–95.

Llewellyn, S., R. Proctor, G. Harvey, G. Maniatopoulos, and A. Boyd. 2014. "Facilitating Technology Adoption in the NHS: Negotiating the Organisational and Policy Context - a Qualitative Study." *Health Services and Delivery Research* 2(23). doi: 10.3310/hsdr02230.

Lomas, Jonathan. 2007. "The in-Between World of Knowledge Brokering." *BMJ* 334(7585): 129–132. doi: 10.1136/bmj.39038.593380.AE.

Madden, Mary. 2012. "Alienating Evidence Based Medicine vs. Innovative Medical Device Marketing: A Report on the Evidence Debate at a Wounds Conference." *Social Science & Medicine* 74(12): 2046–2052. doi: http://dx.doi.org/10.1016/j.socscimed.2012.02.026.

Manji, H., R. S. Howard, D. H. Miller, N. P. Hirsch, L. Carr, K. Bhatia, N. Quinn, and C. D. Marsden. 1998. "Status Dystonicus: The Syndrome and its Management." *Brain* 121(Pt 2): 243–252.

Mankins, John. 1995. Technology Readiness Levels. Office of Space Access and Technology, NASA.

MarketsandMarkets. 2015. Neuromodulation Market by Technology - Trends & Global Forecast to 2020. Pune, India.

Marks, W. A., J. Honeycutt, F. Acosta, and M. Reed. 2009. "Deep Brain Stimulation for Pediatric Movement Disorders." *Semin Pediatr Neurol* 16(2): 90–98. doi: 10.1016/j.spen.2009.04.001.

Marks, Warren, Laurie Bailey, and Terence D. Sanger. 2016. "PEDiDBS: The Pediatric International Deep Brain Stimulation Registry Project". *European Journal of Paediatric Neurology.* doi: http://dx.doi.org/10.1016/j.ejpn.2016.06.002.

Marsden, C. D., and J. D. Parkes. 1977. "Success And Problems Of Long-Term Levodopa Therapy In Parkinson's Disease." *The Lancet* 309(8007): 345–349. doi: 10.1016/s0140-6736(77)91146-1.

Martin, Paul A. 1999. "Genes as Drugs: The Social Shaping of Gene Therapy and The Reconstruction Of Genetic Disease." *Sociology of Health & Illness* 21(5): 517–538. doi: 10.1111/1467-9566.00171.

Martínez-Martín, P., A. Gil-Nagel, L. Morlán Gracia, J. Balseiro Gómez, J. Martínez-Sarriés, and F. Bermejo. 1994. "Unified Parkinson's Disease Rating Scale Characteristics and Structure." *Movement Disorders* 9(1): 76–83. doi: 10.1002/mds.870090112.

Maseide, P. 2011. "Body Work in Respiratory Physiological Examinations." *Sociol Health Illn* 33(2): 296–307. doi: 10.1111/j.1467-9566.2010.01292.x.

Mashour, G. A., E. E. Walker, and R. L. Martuza. 2005. "Psychosurgery: Past, Present, and Future." *Brain Res Brain Res Rev* 48(3): 409–419. doi: 10.1016/j.brainresrev.2004.09.002.

Mauss, Marcel. 1973. "Techniques of the Body." *Economy and Society* 2(1): 70–88. doi: 10.1080/03085147300000003.

May, Carl. 1992. "Individual Care? Power and Subjectivity in Therapeutic Relationships." *Sociology* 26(4): 589–602. doi: 10.1177/0038038 592026004003.

May, Carl. 2013a. "Agency and Implementation: Understanding the Embedding of Healthcare Innovations in Practice." *Social Science & Medicine* 78: 26–33. doi: http://dx.doi.org/10.1016/j.socscimed.2012.11.021.

May, Carl. 2013b. "Towards a General Theory of Implementation." *Implementation Science* 8(1): 1–14. doi: 10.1186/1748-5908-8-18.

May, Carl, and Tracy Finch. 2009. "Implementing, Embedding, and Integrating Practices: An Outline of Normalization Process Theory." *Sociology* 43(3): 535–554. doi: 10.1177/0038038509103208.

May, Carl, Linda Gask, Theresa Atkinson, Nicola Ellis, Frances Mair, and Aneez Esmail. 2001. "Resisting and Promoting New Technologies in Clinical Practice: The Case of Telepsychiatry." *Social Science & Medicine* 52(12): 1889–1901. doi: http://dx.doi.org/10.1016/S0277-9536(00) 00305-1.

May, C., T. Rapley, T. Moreira, T. Finch, and B. Heaven. 2006. "Technogovernance: Evidence, Subjectivity, and the Clinical Encounter in Primary Care Medicine." *Soc Sci Med* 62(4): 1022–1030. doi: 10.1016/j. socscimed.2005.07.003.

May, C., T. Finch, F. Mair, L. Ballini, C. Dowrick, M. Eccles, L. Gask, A. MacFarlane, E. Murray, and T. Rapley. 2007. "Understanding the Implementation of Complex Interventions in Health Care: The Normalization Process Model." *BMC Health Serv Res* 7: 142.

May, Carl, Frances Mair, Tracy Finch, Anne MacFarlane, Christopher Dowrick, Shaun Treweek, Tim Rapley, Luciana Ballini, Bie Ong, Anne Rogers, Elizabeth Murray, Glyn Elwyn, France Legare, Jane Gunn, and Victor Montori. 2009. "Development of a Theory of Implementation and Integration: Normalization Process Theory." *Implementation Science* 4(1): 29.

Mayberg, H. S., A. M. Lozano, V. Voon, H. E. McNeely, D. Seminowicz, C. Hamani, J. M. Schwalb, and S. H. Kennedy. 2005. "Deep Brain Stimulation for Treatment-Resistant Depression." *Neuron* 45(5): 651–660. doi: 10.1016/j.neuron.2005.02.014.

McConachie, H. R., A. Salt, Y. Chadury, A. McLachlan, and S. Logan. 1999. "How Do Child Development Teams Work? Findings From a UK National Survey." *Child Care Health Dev* 25(2): 157–168.

McGee, Ellen M., and Gerald Maguire. 2007. "Becoming Borg to Become Immortal: Regulating Brain Implant Technologies." *Cambridge Quarterly of Healthcare Ethics* 16(03): 291–302. doi: 10.1017/S0963180107070326.

McShane, Martin. 2013. "Introducing the 'House of Care'." *The Health Foundation Newletter* 25 September 2013.

McWhinney, Ian R. 1989. "The Need for a Transformed Clinical Method." In *Communicating with Medical Patients*, edited by M Stewart and D Roter, 25–42. London: Sage Publications.

Mead, Nicola, and Peter Bower. 2000. "Patient-Centredness: A Conceptual Framework and Review of The Empirical Literature." *Social Science & Medicine* 51(7): 1087–1110. doi: http://dx.doi.org/10.1016/S0277-9536 (00)00098-8.

Medical Services Advisory Committee. 2008. *Deep Brain Stimulation for Dystonia and Essential Tremor*. Canberra: MSAC REport no 1092.

Medtronic. 2014. Annual Report: SEC 10-K Filing for Fiscal Year 2014. Minneapolis MN: Medtronic.

Medtronic. 2015. "Parkinson's Disease: Benefits and Risks – DBS Therapy." Medtronic, accessed 20 October. http://www.medtronic.com/patients/par kinsons-disease/therapy/benefits-and-risks/.

Meloni, Maurizio. 2014. "How Biology Became Social, and What it Means for Social Theory." *The Sociological Review*: n/a-n/a. doi: 10.1111/1467-954X.12151.

Mesman, J. 2008. *Uncertainty in Medical Innovation:Experienced Pioneers in Neonatal Care*. Basingstoke: Palgrave Macmillan.

Mittra, J. 2016. *The New Health Bioeconomy: R&D Policy and Innovation for the Twenty-First Century*. Basingstoke: Palgrave Macmillan US.

Mittra, J., and C. Milne. 2013. "Introduction to Translational Medicine." In *Translational Medicine: The Future of Therapy*, edited by J. Mittra and C Milne, 3–13. Singapore: Pan-Stanford.

Mol, Annemarie. 1999. "Ontological Politics. A Word and Some Questions." *The Sociological Review* 47(S1): 74-89. doi: 10.1111/j.1467-954X.1999.tb03483.x.

Mol, A. 2002. *The Body Multiple: Ontology in Medical Practice*. Durham NJ: Duke University Press.

Mol, A. 2008. *The Logic of Care: Health and the Problem of Patient Choice*. Abingdon: Routledge.

Monbaliu, E., E. Ortibus, F. Roelens, K. Desloovere, J. Deklerck, P. Prinzie, P. De Cock, and H. Feys. 2010. "Rating Scales for Dystonia in Cerebral Palsy: Reliability and Validity." *Developmental Medicine & Child Neurology* 52(6): 570–575. doi: 10.1111/j.1469-8749.2009.03581.x.

Montgomery, Erwin B., and John T. Gale. 2008. "Mechanisms of Action of Deep Brain Stimulation (DBS)." *Neuroscience & Biobehavioral Reviews* 32(3): 388–407.

Moreira, Tiago. 2004. "Coordination and Embodiment in the Operating Room." *Body & Society* 10(1): 109–129. doi: 10.1177/1357034x04042169.

Moreira, Tiago. 2006. "Heterogeneity and Coordination of Blood Pressure in Neurosurgery." *Social Studies of Science* 36(1): 69–97. doi: 10.1177/0306312705053051.

Moreira, Tiago, and Paolo Palladino. 2005. "Between Truth and Hope: On Parkinson's Disease, Neurotransplantation and The Production of the 'Self.'" *History of the Human Sciences* 18(3): 55–82. doi: 10.1177/0952695105059306.

Morlacchi, Piera, and Richard R. Nelson. 2011. "How Medical Practice Evolves: Learning to Treat Failing Hearts with an Implantable Device." *Research Policy* 40(4): 511–525. doi: http://dx.doi.org/10.1016/j.respol.2011.01.001

Morlacchi, Piera, and Richard Nelson. 2016. "The Evolution of the Left Ventricular Assist Device as a Treatment for Heart favour." In *Medical Innovation: Science, technology and practice*, edited by David Consoli, Andrea Mina, Richard Nelson, and Ronnie Ramlogan, 48–68. London: Routledge.

Moutaud, B. 2011. "Are we Receptive to Naturalistic Explanatory Models of our Disease Experience? Applications of Deep Brain Stimulation to Obsessive Compulsive Disorder and Parkinson's Disease." In *Sociological Reflections on the Neurosciences*, edited by Martyn Pickersgill and I. Van Keulen, 179–202. Bingley UK: Emerald Group Publishing.

Moutaud, Baptiste. 2012. "Are we Receptive to Naturalistic Explanatory Models of our Disease Experience? Applications of Deep Brain Stimulation to Obsessive Compulsive Disorders and Parkinson's Disease." In *Sociological Reflections on the Neurosciences*, edited by M. Pickersgill and I. Van Keulen, 179–202. Bingley: Emerald Group Publishing Limited.

Moutaud, Baptiste. 2015. "Neuromodulation Technologies and the Regulation of Forms of Life: Exploring, Treating, Enhancing." *Medical Anthropology*: 1–17. doi: 10.1080/01459740.2015.1055355.

Moutaud, B. 2016. "Neuromodulation Technologies and the Regulation of Forms of Life: Exploring, Treating, Enhancing." *Med Anthropol* 35(1): 90–103. doi: 10.1080/01459740.2015.1055355.

Mundinger, F. 1965. "Stereotaxic Interventions on the Zona Incerta Area for Treatment of Extrapyramidal Motor Disturbances and Their Results." *Confin Neurol* 26(3): 222–230.

Mundinger, F. 1977. "[New Stereotactic Treatment of Spasmodic Torticollis with a Brain Stimulation System (Author's Transl)]." *Med Klin* 72(46): 1982–1986.

Nashold Jr., B. S., and D. G. Slaughter. 1969. "Effects of Stimulating or Destroying the Deep Cerebellar Regions in Man." *J Neurosurg* 31(2): 172–186. doi: 10.3171/jns.1969.31.2.0172.

NHS. 2013. *The NHS Constitution: The NHS Belongs to Us All*, edited by The Department of Health. London: Crown.

NHS England. 2014. *2014/15 National Tariff Payment System*. London: The Crown.

NHS National Cancer Action Team. 2010. *The Characteristics of an Effective Multidisciplinary Team (MDT)*. London: NHS.

NHSE. 2016. "House of Care Model - Background." NHS England, accessed 29 July 2016. https://www.england.nhs.uk/resources/resources-for-ccgs/out-frwrk/dom-2/house-of-care/house-care-mod/.

NICE. 2006. *Deep Brain Stimulation for Tremor and Dystonia (Excluding Parkinson's Disease)*. London: Department of Health.

Novas, Carlos. 2006. "The Political Economy of Hope: Patients & Organizations, Science and Biovalue." *BioSocieties* 1(03): 289–305.

Nuffield Council on Bioethics. 2013. Novel Neurotechnologies: Intervening in the Brain. Nuffield Council on Bioethics.

Nygard, L., B. Bernspang, A. G. Fisher, and B. Winblad. 1994. "Comparing Motor and Process Ability of Persons with Suspected Dementia in Home and Clinic Settings." *Am J Occup Ther* 48(8): 689–696.

Ortega, Francisco. 2009. "The Cerebral Subject and the Challenge of Neurodiversity." *BioSocieties* 4(04): 425–445.

Ostrem, Jill L., and Philip A. Starr. 2008. "Treatment of Dystonia with Deep Brain Stimulation." *Neurotherapeutics* 5(2): 320–330.

Oudshoorn, Nelly. 2015. "Sustaining Cyborgs: Sensing and Tuning Agencies of Pacemakers and Implantable Cardioverter Defibrillators." *Social Studies of Science* 45(1): 56–76. doi: 10.1177/0306312714557377.

Pasveer, Bernike. 1989. "Knowledge of Shadows: The Introduction of X-Ray Images in Medicine." *Sociology of Health & Illness* 11(4): 360–381. doi: 10.1111/1467-9566.ep11373066.

Penfield, W. 1936. "Epilepsy and Surgical Therapy." *Archives of Neurology & Psychiatry* 36(3): 449–484. doi: 10.1001/archneurpsyc.1936.02260090002001.

Petryna, Adriana. 2006. "Globalizing Human Subjects Research." In *Global Pharmaceuticals: Ethics, Markets, Practices*, edited by Adriana Petryna, Andrew Lakoff, and Arthure Kleinman, 33–60. Durham: Duke University Press.

Pickersgill, Martyn. 2009. "Between Soma and Society: Neuroscience and the Ontology of Psychopathy." *BioSocieties* 4(1): 45–60. doi: 10.1017/s1745855209006425.

Pickersgill, M. 2011. "'Promising' Therapies: Neuroscience, Clinical Practice, and the Treatment of Psychopathy." *Sociol Health Illn* 33(3): 448–464. doi: 10.1111/j.1467-9566.2010.01286.x.

Pickersgill, M., and I. Van Keulen. 2011. *Sociological Reflections on the Neurosciences: Social and Political Approaches*. Bingley: Emerald Publishing Group Limited.

Pickersgill, Martyn, Sarah Cunningham-Burley, and Paul Martin. 2011. "Constituting Neurologic Subjects: Neuroscience, Subjectivity and the Mundane Significance of the Brain." *Critical Psychology* 4(3): 346–365. doi: 10.1057/sub.2011.10.

Pickstone, J. V. 2001. *Ways of Knowing: A New History of Science, Technology, and Medicine*. Chicago: University of Chicago Press.

Pietroni, Patrick C. 1992. "Towards Reflective Practice—The Languages of Health and Social Care." *Journal of Interprofessional Care* 6(1): 7–16. doi: 10.3109/13561829209049590.

Porras, Gregory, Li Qin, and Erwan Bezard. 2012. "Modeling Parkinson's Disease in Primates: The MPTP Model." *Cold Spring Harbor Perspectives in Medicine* 2(3): a009308. doi: 10.1101/cshperspect.a009308.

Porter, Theodore. 1994. "Making Things Quantitative." *Science in Context* 7(3): 389–407.

Porter, Theodore. 1995. *Trust in Numbers: The Pursuit of Objectivity in Science and Public Life*. Princeton, NJ: Princeton University Press.

Prentice, Rachel. 2007. "Drilling Surgeons: The Social Lessons of Embodied Surgical Learning." *Science, Technology & Human Values* 32(5): 534–553. doi: 10.1177/0895904805303201.

Prior, Lindsay. 1988. "The Architecture of the Hospital: A Study of Spatial Organization and Medical Knowledge." *The British Journal of Sociology* 39(1): 86–113. doi: 10.2307/590995.

Rabinow, Paul. ed. 1991. *The Foucault Reader: An Introduction to Foucault's Thought.* London: Penguin.

Rabinow, Paul. 2008. "Artificiality and Enlightenment: From Sociobiology to Biosociality." In *Anthropologies of Modernity*, edited by P. Rabinow, 179–193. Hoboken NJ: Blackwell Publishing Ltd.

Racine, Eric, Sarah Waldman, Nicole Palmour, David Risse, and Judy Illes. 2007. ""Currents of Hope": Neurostimulation Techniques in the U.S. and U.K. Print Media." *Cambridge Quarterly of Healthcare Ethics* 16: 312–316.

Rapp, Rayna. 2012. "A Child Surrounds this Brain: The Future of Neurological Difference According to Scientists, Parents and Diagnosed Young Adults." In *Sociological Reflections on the Neurosciences*, 3–26. Bingley: Emerald Group Publishing Limited.

Reeves, A., and R. Swenson. 2008. *Disorders of the Nervous System.* Hanover NH: Dartmouth Medical School.

Regenerative Medicine Expert Group. 2015. *Building on our own Potential: A UK Pathway for Regenerative Medicine.* London: Department of Health.

Ritzer, G. 2014. *The McDonaldization of Society.* Thousand Oaks CA: SAGE Publications.

Rose, Nikolas. 2001. "The Politics of Life Itself." *Theory, Culture & Society* 18(6): 1–30. doi: 10.1177/02632760122052020.

Rose, N. 2003. "Neurochemical Selves." *Society* 41: 46–59.

Rose, Nikolas. 2007. *The Politics of Life Itself: Biomedicine, Power, and Subjectivity in the Twenty-First Century.* Princeton and Oxford: Princeton University Press.

Rose, N., and J. M. Abi-Rached. 2013. *Neuro: The New Brain Sciences and the Management of the Mind.* Princeton: Princeton University Press.

Rose, N., and C. Novas. 2004. "Biological Citizenship." In *Global Assemblages: Technology, Politics and Ethics as Anthropological Problems*, edited by A. Ong and S. Collier, 439–463. New York: Blackwell.

Rosenow, J., K. Das, R. L. Rovit, and W. T. Couldwell. 2002. "Irving S. Cooper and His Role in Intracranial Stimulation for Movement Disorders and Epilepsy." *Stereotactic and Functional Neurosurgery* 78(2): 95–112.

Rossi, U. 2003. "The History of Electrical Stimulation of the Nervous System for the Control of Pain." In *Electrical Stimulation and the Relief of Pain*, edited by B. Simpson, 5–16. Amsterdam: Elsevier.

Russell, D. J., P. L. Rosenbaum, D. T. Cadman, C. Gowland, S. Hardy, and S. Jarvis. 1989. "The Gross Motor Function Measure: A Means to Evaluate the Effects of Physical Therapy." *Dev Med Child Neurol* 31(3): 341–352.

Ryder, D. 2011. "Subjective Examination." In *Neuromuscularskeletal Examination and Assessment: A Handbook for Therapists*, edited by N Petty. London: Churchill Livingstone Elsevier.

Schlaepfer, Thomas E., and Joseph J. Fins. 2010. "Deep Brain Stimulation and the Neuroethics of Responsible Publishing." *JAMA: The Journal of the American Medical Association* 303(8): 775–776. doi: 10.1001/jama.2010.140.

Schot, Johan, and Frank W. Geels. 2007. "Niches in Evolutionary Theories of Technical Change." *Journal of Evolutionary Economics* 17(5): 605–622. doi: 10.1007/s00191-007-0057-5.

Schubert, Cornelius. 2011. "Making Sure. A Comparative Micro-Analysis of Diagnostic Instruments in Medical Practice." *Social Science & Medicine* 73(6): 851–857. doi: http://dx.doi.org/10.1016/j.socscimed.2011.05.032.

Schüpbach, M., M. Gargiulo, M. L. Welter, L. Mallet, C. Béhar, J. L. Houeto, D. Maltête, V. Mesnage, and Y. Agid. 2006. "Neurosurgery in Parkinson Disease: A Distressed Mind in a Repaired Body?" *Neurology* 66(12): 1811–1816. doi: 10.1212/01.wnl.0000234880.51322.16.

Shatin, Deborah, Keith Mullett, and Gerald Hults. 1986. "Totally Implantable Spinal Cord Stimulation for Chronic Pain: Design and Efficacy." *Pacing and Clinical Electrophysiology* 9(4): 577–583. doi: 10.1111/j.1540-8159.1986.tb06614.x.

Shealy, C. N., J. T. Mortimer, and J. B. Reswick. 1967. "Electrical Inhibition of Pain by Stimulation of the Dorsal Columns: Preliminary Clinical Report." *Anesth Analg* 46(4): 489–491.

Siegfried, Jean, and Joseph Shulman. 1987. "SESSION II: Deep Brain Stimulation." *Pacing and Clinical Electrophysiology* 10(1): 271–272. doi: 10.1111/j.1540-8159.1987.tb05961.x.

Simon, Herbert. 1979. "Rational Decision Making in Business Organizations." *American Economic Review* 69(4): 493–513.

Singh, I. 2013. "Brain Talk: Power and Negotiation in Children's Discourse About Self, Brain and Behaviour." *Sociol Health Illn* 35(6): 813–827. doi: 10.1111/j.1467-9566.2012.01531.x.

Sironi, Vittorio A. 2011. "Origin and Evolution of Deep Brain Stimulation." *Frontiers in Integrative Neuroscience* 5: 42. doi: 10.3389/fnint.2011.00042.

Smith, Robert C. 2002. "The Biopsychosocial Revolution: Interviewing and Provider-Patient Relationships Becoming Key Issues for Primary Care." *Journal of General Internal Medicine* 17(4): 309–310. doi: 10.1046/j.1525-1497.2002.20210.x.

Spiegel, E. A., H. T. Wycis, M. Marks, and A. J. Lee. 1947. "Stereotaxic Apparatus for Operations on the Human Brain." *Science* 106(2754): 349–350. doi: 10.1126/science.106.2754.349.

Star, Susan Leigh, and Anselm Strauss. 1999. "Layers of Silence, Arenas of Voice: The Ecology of Visible and Invisible Work." *Computer Supported Cooperative Work (CSCW)* 8(1): 9–30. doi: 10.1023/a:1008651105359.

Stuart, M. 2012. "Deep Brain Provides Stimulating Market." *Medtech Insight* 14: 4.

Synofzik, Matthis, and Thomas E. Schlaepfer. 2011. "Electrodes in the Brain– Ethical Criteria for Research and Treatment with Deep Brain Stimulation for Neuropsychiatric Disorders." *Brain Stimulation* 4(1): 7–16.

Talan, Jamie. 2009. *Deep Brain Stimulation: A New Treatment Shows Promise in the Most Difficult Cases.* New York: Dana Press.

Thompson, A. G. 2007. "The Meaning of Patient Involvement and Participation in Health Care Consultations: A Taxonomy." *Soc Sci Med* 64(6): 1297–1310. doi: 10.1016/j.socscimed.2006.11.002.

Tieman, B. L., R. J. Palisano, and A. C. Sutlive. 2005. "Assessment of Motor Development and Function in Preschool Children." *Ment Retard Dev Disabil Res Rev* 11(3): 189–196. doi: 10.1002/mrdd.20074.

Tomlin, Zelda, Susan Peirce, Glyn Elwyn, and Alex Faulkner. 2013. *The Adoption Space of Early-Emerging Technologies: Evaluation, Innovation, Gatekeeping (PATH). Final report.* Edited by NIHR Service Delivery and Organisation programme.

Tutton, Richard. 2011. "Promising Pessimism: Reading the Futures to be Avoided in Biotech." *Social Studies of Science* 41(3): 411–429. doi: 10.1177/0306312710397398.

UK Research Councils. 2012. *A Strategy for UK Regenerative Medicine.* London: Medical Research Council.

Ulucanlar, S., A. Faulkner, S. Peirce, and G. Elwyn. 2013. "Technology Identity: The Role of Sociotechnical Representations in the Adoption of Medical Devices." *Social Science & Medicine* 98: 95–105. doi: http://dx.doi.org/10.1016/j.socscimed.2013.09.008.

Upton, Adrian. 1986. "Biostimulation." *Pacing and Clinical Electrophysiology* 9(1): 3–7. doi: 10.1111/j.1540-8159.1986.tb05354.x.

Upton, A., and Yves Lazorthes. 1987. "Editorial." *Pacing and Clinical Electrophysiology* 10:161.

Valenstein, E. 1997. "The History of Pschosurgery in its Scientific and Professional Contexts." In *The History of Neurosurgery*, edited by S Greenblatt, 499–516. Park Ridge IL.: AANS.

Van Zelst, B. R., M. D. Miller, R. Russo, S. Murchland, and M. Crotty. 2006. "Activities of Daily Living in Children with Hemiplegic Cerebral Palsy: A Cross-Sectional Evaluation Using the Assessment of Motor and Process Skills." *Dev Med Child Neurol* 48(9): 723–727. doi: 10.1017/s0012162206001551.

Vercueil, Laurent, Pierre Pollak, Valérie Fraix, Elena Caputo, Elena Moro, Abdelhamid Benazzouz, Jing Xie, Adnan Koudsie, and Alim-Louis Benabid. 2001. "Deep Brain Stimulation in the Treatment of Severe Dystonia." *Journal of Neurology* 248(8): 695–700. doi: 10.1007/s004150170116.

Vidal, Fernando. 2009. "Brainhood, Anthropological Figure of Modernity." *History of the Human Sciences* 22(1): 5–36. doi: 10.1177/0952695108099133.

Webster, Andrew. 2002. "Innovative Health Technologies and the Social: Redefining Health, Medicine and the Body." *Current Sociology* 50(3): 443–457. doi: 10.1177/0011392102050003009.

Wehrens, R., and R. Bal. 2012. "Health Programs Struggling with Complexity: A Case Study of the Dutch 'Precare' Project." *Soc Sci Med* 75(2): 274–282. doi: 10.1016/j.socscimed.2012.02.052.

West, M., and J. Slater. 1996. *The Effectiveness of Team Working in Primary Health Care*. London: Health Education Authority.

Yarnold, J. 2009. "Early and Locally Advanced Breast Cancer: Diagnosis and Treatment National Institute for Health and Clinical Excellence Guideline 2009." *Clinical Oncology* 21(3): 159–160. doi: http://dx.doi.org/10.1016/j.clon.2008.12.008.

Yianni, J., A. L. Green, E. McIntosh, R. G. Bittar, C. Joint, R. Scott, R. Gregory, P. G. Bain, and T. Z. Aziz. 2005. "The Costs and Benefits of Deep Brain Stimulation Surgery for Patients with Dystonia: An Initial Exploration." *Neuromodulation* 8(3): 155–161. doi: 10.1111/j.1525-1403.2005.05233.x.

Zeisel, J. 2006. *Inquiry by Design: Environment/Behavior/Neuroscience in Architecture, Interiors, Landscape, and Planning*. New York: W.W. Norton.

Zemel, Alan, and Timothy Koschmann. 2014. "'Put Your Fingers Right in Here': Learnability and Instructed Experience." *Discourse Studies* 16(2): 163–183. doi: 10.1177/1461445613515359.

Index

© The Author(s) 2017
J. Gardner, *Rethinking the Clinical Gaze*, Health, Technology
and Society, DOI 10.1007/978-3-319-53270-7